GEAR DRIVE SYSTEMS
DESIGN AND APPLICATION

MECHANICAL ENGINEERING

A Series of Textbooks and Reference Books

EDITORS

L. L. FAULKNER

Department of Mechanical Engineering
The Ohio State University
Columbus, Ohio

S. B. MENKES

Department of Mechanical Engineering
The City College of the
City University of New York
New York, New York

OTHER VOLUMES IN PREPARATION

GEAR DRIVE SYSTEMS

DESIGN AND APPLICATION

Peter Lynwander
American Lohmann Corporation
Hillside, New Jersey

MARCEL DEKKER, INC.   New York and Basel

Library of Congress Cataloging in Publication Data

Lynwander, Peter, [date]
  Gear drive systems.

  (Mechanical engineering ; 20)
  Includes index.
  1. Gearing. I. Title. II. Series.
TJ184.L94 1983   621.8'33   83-5278
ISBN 0-8247-1896-8

MARCEL DEKKER, INC.
270 Madison Avenue, New York, New York 10016

Current printing (last digit):
10 9 8 7 6 5 4 3 2 1

PRINTED IN THE UNITED STATES OF AMERICA

PREFACE

Gear drives are critical components of mechanical systems used in such diverse industries as turbo-machinery, process, refinery, steel, construction, mining, and marine. In all these fields there is a continuing trend toward higher reliability and improved technology in mechanical components. Higher reliability is desired to reduce downtime. In many applications, the cost of one day's lost production due to a gearbox malfunction far exceeds the initial cost of the unit; therefore, in critical installations there is a strong emphasis on conservative design and quality manufacture. In addition to achieving high reliability, mechanical systems must be increasingly efficient to conserve energy. Gear manufacturers are constantly refining their analytical, design, and manufacturing techniques to take advantage of new technologies and provide reliable, efficient gearboxes at minimum cost.

The purpose of this book is to present practical gearbox design and application information to individuals responsible for the specification and operation of mechanical systems incorporating gear drives. Sufficient theoretical information is included to enable the engineer interested in gear analysis and design to understand how gear units are rated and detail gear tooth geometry is defined. The major emphasis is on parallel shaft and planetary units using spur and helical gearing.

In addition to basic data on gear design and manufacture, such subjects as installation, operation, maintenance, troubleshooting, failure analysis, and economics are covered. Material on lubrication systems, bearings, couplings, and seals is presented in order to cover all aspects of gear system operation.

Several new trends in the gear industry, due in part to the emphasis on energy conservation, are discussed.

1. As mechanical equipment such as pumps, motors, compressors, turbines, etc. are designed for higher efficiencies, rotating speeds are increased and, therefore, higher speed transmissions are required. High speed gearing characteristics are featured throughout the book.

2. Also, as a result of energy consciousness, there is a tendency to package smaller mechanical systems; therefore, there is a trend developing in the United States toward the use of planetary gear units which are far more compact than parallel shaft designs. Included in the book is a section on planetary gear design and application.

3. In order to achieve the highest load carrying capability in a minimum envelope, case hardened and ground gear tooth designs are finding wide application. This technology is covered in the book.

4. The book attempts to take a systems approach to gearbox application. It has become apparent that gear units, when incorporated into a system of rotating machinery, are susceptible to a variety of problems. All characteristics of the drive system from the driver to the driven equipment, including the lubrication system and accessories, can influence gearbox operation and must be considered in the specification, installation, operation, and maintenance of the unit.

Throughout the book, standards and practices developed by the American Gear Manufacturers Association are referred to. Successful selection, rating, and operation of gearboxes can be accomplished by the use of AGMA publications and the gear designer and user should be familiar with the Standards system. The AGMA is located at 1901 North Fort Myer Drive, Arlington, Virginia 22209.

I would like to thank Mr. Alvin Meyer and Mr. Alan Swirnow for their assistance and comments. I am also indebted to American Lohmann Corporation, Hillside, New Jersey for its support.

Peter Lynwander

CONTENTS

GEAR DRIVE SYSTEMS

DESIGN AND APPLICATION

1
TYPES OF GEAR DRIVES: ARRANGEMENTS, TOOTH FORMS

The function of a gearbox is to transmit rotational motion from a driving prime mover to a driven machine. The driving and driven equipment may operate at different speeds, requiring a speed-increasing or speed-decreasing unit. The gearbox therefore allows both machines to operate at their most efficient speeds. Gearboxes are also used to change the sense of rotation or bridge an angle between driving and driven machinery.

The gearbox configuration chosen for a given application is most strongly influenced by three parameters:

Physical arrangement of the machinery
Ratio required between input and output speeds
Torque loading (combination of horsepower and speed)

Other factors that must be considered when specifying a gear drive are:

Efficiency
Space and weight limitations
Physical environment

PHYSICAL ARRANGEMENT

The location of the driving and driven equipment in the mechanical system defines the input and output shaft geometrical relationship. Shaft arrangements can be parallel offset, concentric, right angle, or skewed as shown in Figure 1.1. The material presented in this book focuses on parallel offset and concentric designs.

1

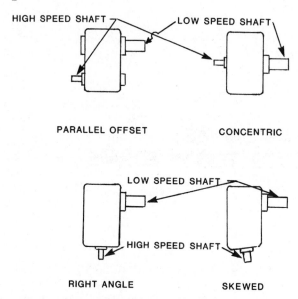

Figure 1.1 Gearbox shaft arrangements.

In the majority of parallel offset units in use, the input and output shafts are horizontally offset; however, vertical offsets are used and any orientation of input to output shaft is possible. Figure 1.2 illustrates a typical horizontally offset parallel shaft gearbox and Figure 1.3 presents a cutaway view of such a unit. In this case there is one input shaft and one output shaft located on opposite sides of the unit. There are many different options available as far as the input and output shaft extensions are concerned. Figure 1.4 shows the various possibilities and presents a system for defining the extensions desired on a gearbox. There may be two inputs driving a single output, such as dual turbines powering a large generator, or two outputs with a single input, such as an electric motor driving a two-stage compressor. Often shaft extensions are used to drive accessories such as pumps or starters.

The minimum amount of offset required is determined by gear tooth stress considerations. The offset of a gearbox incorporating a single mesh, as shown in Figure 1.3, is the sum of the pitch radii of the pinion and gear, otherwise known as the center distance. The pitch radii must be sufficiently large to transmit the system load. An offset greater than the minimum may be required to provide enough space for the machinery incorporated in the system. Figure 1.5 illustrates an accessory drive where the input and output shafts are offset through two meshes to separate the machinery located at these shafts.

Figure 1.2 Parallel offset gearbox. (Courtesy of American Lohmann Corporation, Hillside, N.J.)

Figure 1.6 illustrates a gearbox with concentric input and output shafts. The driving and driven machinery will therefore be in line. Planetary gearing (described in the next section) has concentric shafts and is used to achieve high ratios in minimum space. It is also possible to package parallel shaft gearing such that the input and output shafts are in line when such a configuration is desired. Figure 1.7 presents external and internal views of a right-angle gear drive.

The gearboxes in Figures 1.3, 1.6, and 1.7 are foot mounted; that is, they are meant to be bolted to a horizontal base through a flange at the bottom of

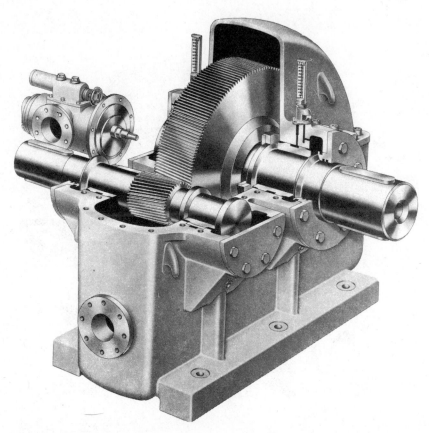

Figure 1.3 Parallel offset gearbox sectional view. (Courtesy of American Lohmann Corporation, Hillside, N.J.)

the gear casing. Although this is the most common design, gearboxes can be mounted in many other configurations and operate in attitudes other than horizontal. Figure 1.8 illustrates a flange-mounted unit. Such a gearbox can be operated horizontally or be vertically mounted on a horizontal base. Vertically operating gearbox designs must have special lubrication provisions to provide lubricant to the upper components in the unit and seal the lower end from oil leakage. Figure 1.9 shows yet another mounting configuration. This unit is shaft mounted with a support arm that is fixed to ground to react the gearbox housing torque.

PLAN VIEWS

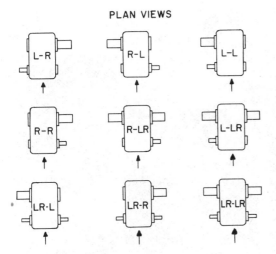

Figure 1.4 Definition of shaft extensions. Parallel shaft-helical and herringbone gear reducers; single, double, and triple reduction. Code; L = left; R = right; arrows indicate line of sight to determine direction of shaft extensions; letters preceding the hyphen refer to number and direction of highspeed shaft extensions; letters following the hyphen refer to number and direction of low speed shaft extensions. (From Ref. 1.)

Figure 1.5 Parallel offset accessory drive gearbox. (Courtesy of American Lohmann Corporation, Hillside, N.J.)

Figure 1.6 Gearbox with concentric input and output shafts. (Courtesy of American Lohmann Corporation, Hillside, N.J.)

GEAR RATIO

There is no limit to the reduction or speed increasing ratio that can be achieved using gearing; however, for high ratios the arrangement of the components can be quite complex. In a simple gear mesh a maximum ratio in the order of 8:1 to 10:1 can be achieved. The amount of speed reduction or increase is simply the ratio of the pitch diameter of the larger gear to the smaller gear. The number of teeth in a gear pair is related to the pitch diameters, so the speed ratio can also be calculated by dividing the larger number of teeth by the smaller. The smaller gear is often called the pinion. To attain a ratio of 10:1, therefore, the gear must be 10 times larger than the pinion and there usually are stress or geometrical limitations on the pinion when this ratio is exceeded. To achieve higher ratios with parallel shaft gearing, stages of meshes are combined as shown in Figure 1.10. This unit has three stages of reduction and achieves ratios on the order of 100:1.

    An efficient method of achieving high reduction ratios in minimum space is the use of planetary gearing. This design, completely described in Chapter 9, is illustrated in Figure 1.11. The high-speed sun gear meshes with a number of

Figure 1.7 Right-angle gear drive. (Courtesy of American Lohmann Corporation, Hillside, N.J.)

Figure 1.8 Flange-mounted gear unit. (Courtesy of American Lohmann Corporation, Hillside, N.J.)

Figure 1.9 Shaft-mounted gearbox. (Courtesy of American Lohmann Corporation, Hillside, N.J.)

Figure 1.10 Multi-stage parallel shaft gearbox. (Courtesy of American Lohmann Corporation, Hillside, N.J.)

planets, usually three, which in turn mesh with a ring gear. The ring gear has internal teeth. Either the ring gear or the planet carrier rotate at the low speed of the gear set. Occasionally, all three members are connected to rotating equipment. When the low-speed shaft is either the ring gear or the planet carrier, the ratios in a planetary gearset are:

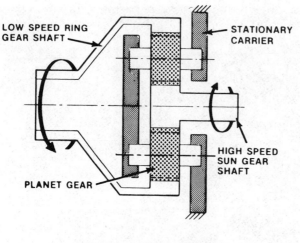

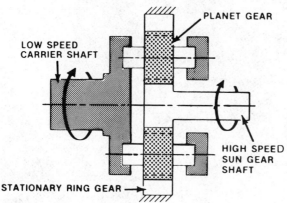

Figure 1.11 Basic planetary gear configurations.

$$\text{Ratio} = \frac{\text{ring gear pitch diameter}}{\text{sun gear pitch diameter}} \quad \text{for rotating ring gear}$$

$$\text{Ratio} = 1 + \frac{\text{ring gear pitch diameter}}{\text{sun gear pitch diameter}} \quad \text{for rotating planet carrier}$$

Because of the multiple load path of planetary gearing the horsepower transmitted is divided between several planet meshes and the gear size can be reduced significantly compared to parallel shaft designs. Planetary stages can be linked together to achieve high ratios, as shown in Figure 1.12. This is a three-stage planetary gear with a ratio of 630:1. The first-stage planet carrier drives the second-stage sun gear and the second-stage carrier drives the third-stage sun gear.

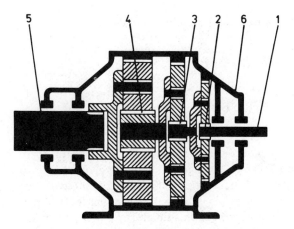

Figure 1.12 Multistage planetary gearbox. High-speed shaft, 1; first planetary stage, 2; second stage, 3; third stage, 4; low-speed shaft, 5; gear housing, 6.

In theory, any parallel shaft or planetary gearbox can be used either as a speed reducer or increaser. There may be details within a gearbox, however, that require modification if such a changeover is made. The same holds true if it is desired to reverse the direction of rotation for which the gearbox was initially designed. For instance, one side of the teeth may have been favored in the finishing process when the initial design was manufactured. If the gearbox is used in such a manner that the initially unloaded face is now loaded, poor tooth performance may result. The direction of rotation of the input shaft with respect to the output shaft depends on the gear design chosen. For a simple parallel shaft gear mesh the sense of rotation will change through the mesh. A planetary arrangement with a stationary ring gear will not change the sense of rotation between input and output, while a rotating ring gear will turn in the opposite sense compared to the sun gear's rotation.

TORQUE LOADING

The size of gearbox required for a given application is dependent primarily on how large the gear pitch diameters and face widths are. These dimensions are determined on the basis of tooth stresses which are imposed by the transmitted tooth load. The tooth load is simply the torque on a given gear divided by the gear pitch radius:

$$\text{Tooth load (lb)} = \frac{\text{torque (in.-lb)}}{\text{pitch radius (in.)}}$$

Torque is calculated from the horsepower transmitted and the speed of the rotating component in question:

$$\text{Input torque (in.-lb)} = \frac{63,025 \text{ (hp)}}{\text{input rpm}}$$

$$\text{Output torque (in.-lb)} = \frac{63,025 \text{ (hp)}}{\text{output rpm}}$$

When designing a gearset one cannot consider torque alone. The operating speed of the gears has a significant effect on the design definition. As an illustration of this point, consider a high-speed unit transmitting 2000 hp at 20,000 rpm input. The input torque would be the same as a low-speed unit operating at 2000 rpm input with a transmitted horsepower of 200. On a simple stress basis, if the ratio of both gearboxes were the same, the same gearbox could be used for both applications; however, the high-speed design must differ from the low-speed design in the following respects:

At high speeds, component geometry discrepancies such as tooth spacing error, shafting unbalance, and so on, generate significant dynamic loading, and these dynamic effects must be taken into account in the design process. Also, the components experience high numbers of load cycles and are more prone than low-speed units to fatigue failures. For all of these reasons, high-speed components must be of high accuracy to minimize dynamic problems.

Heat generation within the unit is proportional to speed; therefore, high-speed units usually require pressure jet lubrication systems and external cooling systems. Low-speed units often operate with integral splash lubrication, the heat being dissipated through the gear casing.

The bearing design is strongly dependent on shaft speeds. Low-speed units generally incorporate antifriction bearings, while high-speed industrial gearboxes typically use journal bearing designs.

There is no clear demarcation between low-speed and high-speed gearing. Units with several gear meshes may have some of each. An arbitrary definition sometimes used is that units with pinion speeds exceeding 3600 rpm or pitch line velocities exceeding 5000 fpm are considered high speed [1].

Pitch line velocity is a measure of the peripheral speed of a gear:

$$\text{Pitch line velocity (fpm)} = \frac{\pi \text{ (pitch diameter, in.) (rpm)}}{12}$$

The pitch line velocity of a gear is a better index of speed than is rotational velocity, since a large gear operating at a relatively low rpm may experience the same velocity effects as a small gear operating at high rpm. Standard high-speed gear units operate at pitch line velocities up to approximately 20,000 fpm.

Applications exceeding this speed must be considered special and exceptional care must be taken in their design and manufacture. Pitch line velocities of 40,000 fpm have been attained in practice.

Parallel offset or concentric shaft gearboxes incorporate gears with spur, single helical, or double helical tooth forms. The face of a spur gear is parallel to the axis of rotation, whereas a helical gear tooth face is at an angle, as shown in Figure 1.13. The figure illustrates that helical gears have an overlap in the axial direction, which results in the following advantages:

Helical gears have more face width in contact than do spur gears of the same size; therefore, they have greater load-carrying capability.

With conventional spur gearing the load is transmitted by either one or two teeth at any instant; thus the elastic flexibility is continuously changing as load is transferred from single-tooth to double-tooth contact and back. With helical gearing the load is shared between sufficient teeth to allow a smoother transference and a more constant elastic flexibility; therefore, helical gearing generates less noise and vibration than spur gearing.

The disadvantage of helical gearing in relation to spur gearing is that axial thrust is generated in a helical gear, which necessitates the incorporation of a thrust bearing on each helical gear shaft.

To take advantage of the helical gearing benefits described above, yet not generate axial thrust loads, double helical gearing is used (Figure 1.14). The two halves generate opposite thrust loads, which cancel out. When the two helices are cut adjacent to one another with no gap between, the gearing is termed herringbone. Because helical gear thrust is proportional to the tangent of the helix angle, single helical gears tend to have lower helix angles than do double helical designs, where the thrust loads cancel. Typical single helical helix angles are 6 to 15°. Double helical gearsets have helix angles of up to 35°.

Another advantage of double helical gears is that the ratio of face width to pitch diameter in each half can be held to reasonable limits. When the face widths become longer than the pitch diameters in spur or single helical gearing it is difficult to achieve complete tooth contact since thermal distortion, load deflections, and manufacturing errors tend to load the gear teeth unevenly. A double helical gear with a face width/pitch diameter ratio of 1 will have twice the face width of a spur or single helical gear with the same L/D ratio and therefore greater load-carrying capability.

Double helical gearing has two disadvantages. Because the two halves of each gear cannot be perfectly matched, one member of the gearset must be free to float axially. This gear will be continually shifting to achieve axial force equilibrium since the thrust loads of each half will rarely cancel exactly. This shifting can lead to detrimental axial vibrations if tooth geometry errors are excessive. Another potential problem with double helical gearing is that external

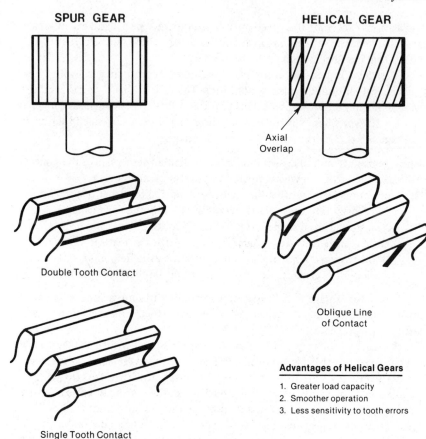

SPUR GEAR

HELICAL GEAR

Axial
Overlap

Double Tooth Contact

Oblique Line
of Contact

Advantages of Helical Gears

1. Greater load capacity
2. Smoother operation
3. Less sensitivity to tooth errors

Single Tooth Contact

Figure 1.13   Comparison of spur and helical gear teeth.

thrust loads will tend to overload one helix. For instance, if a double helical
gear is attached to a gear tooth type of coupling and the coupling locks up
axially due to tooth friction, axial loads transmitted through the coupling will
be reacted by the teeth of one-half of the gearset. With single helical gears
an external axial load will either add to or subtract from the gear tooth load and
be reacted by the thrust bearing.

Gear metallurgy, although not mentioned heretofore, is one of the most
significant factors in determining gearbox size, since the strength of a gear tooth
is proportional to the hardness of the steel. Most gears are in the hardness ranges
of approximately Rc 30 to 38 or Rc 55 to 64. The region from Rc 30 to 38 is
usually termed "through-hardened," while the range Rc 55 to 64 is almost
always "surface-hardened," where the tooth has a hard surface case and a softer
inner core. Through-hardened gears are cut by such processes as hobbing,

Figure 1.14 Double helical gearing.

shaping, and shaving. Surface-hardened gears are cut and then hardened. They may be used in this state, but the more accurate surface-hardened gears are ground after heat treatment. Spur, single helical, and double helical gearing may be produced by any of the methods noted above. Generally, double helical gearing is through-hardened and cut. It is possible to harden and grind double helical gearing; however, to grind a one-piece double helical gear a large central gap is required between the two helices to allow runout of the grinding wheel. Gears can be ground in halves and then assembled, but this presents serious alignment and attachment problems.

To achieve minimum envelope and maximum reliability, the latest technology utilizes single helical, hardened, and precision ground gearing. With single helical gears the thrust load axially locates the gear shaft against the thrust bearing. Bearing design has progressed to the point where thrust loads are routinely handled either by hydrodynamic tapered land or tilting pad configurations or an antifriction thrust bearings. Because case-hardened gears have maximum

Figure 1.15 High-speed single helical hardened and ground gearset. (Courtesy of American Lohmann Corporation, Hillside, N.J.)

load-carrying capacity, gear size can be minimized; therefore, the ratio of face width to diameter of a single helical gear can be held to reasonable limits. Pitch line velocities are minimized, reducing dynamic effects. Also, the bearing span with single helical gears is short, resulting in lower elastic deflection. Figure 1.15 illustrates a generator drive gearbox with two stages of single helical gearing. This unit transmits 4500 hp at an input speed of 14,500 rpm. The high-speed mesh pitch line velocity is 18,000 fpm. Gearbox weight is 3500 lb.

A single helical hardened and ground gearset can reduce by up to one-half the envelope and weight of a through-hardened double helical gearbox with equivalent capacity. The inherent precision of the grinding process results in accurate tooth geometry, leading to minimum noise and vibration.

EFFICIENCY

Gearbox efficiency is a much discussed subject, but accurate values are very difficult to determine. Analytical estimates must be confirmed by testing to gain a degree of confidence in the procedure. With good design and manufacturing practice, efficiencies of 99% per mesh and better are possible. Often, lubrication system development is required to attain the highest efficiency potential.

Power losses in a gearbox are divided between friction losses at the gear and bearing contacts and windage losses as the rotating components churn the oil and air. In high-speed units the churning losses may exceed the friction losses; therefore, the type and amount of lubricant, and its introduction and evacuation, are critical in terms of efficiency. Journal bearings require significantly more oil flow than do antifriction bearings and generate higher power losses.

Figure 1.16 Compound planetary gearset. (Courtesy of American Lohmann Corporation, Hillside, N.J.)

A reasonable estimate of efficiency for industrial gear boxes is 1 to 2% power loss per mesh. A three-stage unit, therefore, might be expected to have an efficiency in the range 94 to 97%. The efficiency is quoted at the design load and speed conditions. At full speeds and lower loads the efficiency will drop off because the churning losses will remain constant.

SPACE AND WEIGHT LIMITATIONS

There are industrial applications where gearbox space and weight is limited. For instance, generator-drive gearboxes on offshore oil platforms or units used on mobile equipment must have minimum envelope. To achieve small gear units, several techniques can be used:

To minimize gear size, the highest-quality steel, case carburized and precision ground, is incorporated in the unit.
Planetary configurations are used to achieve high ratios in small envelopes. Figure 1.16 illustrates a compound planetary gearset which demonstrates a very efficient use of space, in the approximate range 9:1 to 12:1.
Lightweight design techniques such as thin-wall casings and hollow shafts are employed.
Lightweight materials such as aluminum housings are used.

Maximum application of these techniques can be found in the aerospace industry. An aircraft gearbox might handle the same design conditions as a conventional unit but at one-fiftieth the weight [2].

PHYSICAL ENVIRONMENT

When specifying a gear drive, the physical environment must be addressed in the design stage. Listed below are detrimental environments which can have an adverse effect on lubricant, bearings, gears, or seals:

Dusty atmosphere
High ambient temperature
Wide temperature variation
High humidity
Chemical-laden atmosphere

Such environments require special consideration in the design of the gearbox lubrication system and seals.

REFERENCES

1. AGMA Standard 420.04, Practice for Enclosed Speed Reducers or Increasers Using Spur, Helical, Herringbone and Spiral Bevel Gears, American Gear Manufacturers Association, Arlington, Va., December 1975.
2. Dudley, D. W., *Gear Handbook*, McGraw-Hill, New York, 1962, pp. 3–5.

2
GEAR TOOTH DESIGN

The purpose of gearing is to transmit power and/or motion from one shaft to another at a constant angular velocity. The tooth form almost universally used is the involute, which has properties that make it particularly desirable for these functions. It will be shown that in order to attain constant angular velocity, the meshing tooth forms must have specific geometrical characteristics which are easily obtained with an involute system.

In order to understand gear tooth drives it is useful to observe the dynamics of simpler power transmission systems, such as friction disks or belt drives (Figure 2.1), both of which are capable of transmitting power at a constant velocity ratio. The velocity ratio is inversely proportional to the ratio of the diameters:

$$\frac{W_A}{W_B} = \frac{D_B}{D_A}$$

where

$\quad$ W = angular velocity, rad/sec
$\quad$ D = diameter, in.

If $D_A = \frac{1}{2}D_B$ it can be seen that friction disk A has to make two revolutions for each revolution of disk B if the circumferences of the disks are rolling on one another without slipping. Another way of looking at it is that at the point of contact both disks have the same tangential velocity V_T in inches per second, and

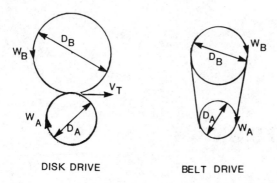

DISK DRIVE   BELT DRIVE

Figure 2.1   Friction disk and belt drives.

$$V_T = \frac{W_A D_A}{2} = \frac{W_B D_B}{2}$$

Therefore,

$$\frac{W_A}{W_B} = \frac{D_B}{D_A}$$

Similarly, the ratio of angular velocity of sheaves A and B in the belt drive are proportional to the ratio of diameter B to diameter A.

Disks and belt drives are power and speed limited and sometimes slip. A more positive method of transmitting power is through gear teeth, which can be illustrated as two cam profiles acting on one another (Figure 2.2). The force of the driving cam on the driven at any instant of time acts normal to the point of tangency of the curved surfaces. This normal line, shown as AA in Figure 2.2, is known as the line of action. Line AA intersects a line drawn between the two centers of rotation at point X. R_A and R_B are then the instantaneous pitch radii of the two cams. The angular velocity ratio of the cams at a given instant is inversely proportional to the ratio of the instantaneous pitch radii. For the angular velocity ratio to remain constant, the respective pitch radii must be the same at all points of contact. If this condition is met, the two profiles are said to be conjugate. Two cam profiles chosen at random will rarely be conjugate; however, given one profile a conjugate mating profile can be developed mathematically. The problem is that these two conjugate profiles may not be practical from an operating or manufacturing point of view.

This leads to one major reason why the involute curve is widely used for gear teeth. Two mating involutes will always be conjugate and the tooth forms relatively easy to manufacture with standardized tooling.

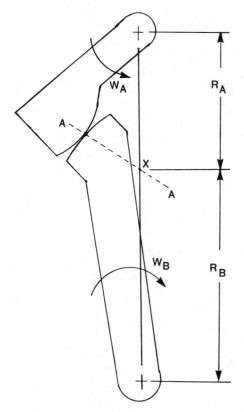

Figure 2.2 Cam drive.

THE INVOLUTE CURVE

Figure 2.3 illustrates the involute curve, which may be visualized as the locus of points generated by the end of a string which is held in tension as it is unwound from a drum. The drum is known as the base circle and once the base circle diameter is known, the involute curve is completely defined. Mathematically, the involute is expressed as a vectorial angle θ in radians:

$$\theta = \tan \phi - \phi = \text{Inv } \phi$$

where ϕ is the pressure angle at any diameter, in radians. In order to plot the involute curve for a base circle of radius R_B, simply assume values for the pressure angle ϕ. All the terms shown in Figure 2.3 can then be calculated.

R is the radius to any point on the involute and is related to R_B by the cosine of the pressure angle:

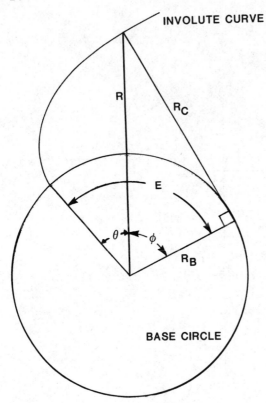

Figure 2.3   The involute curve.

$$\cos \phi = \frac{R_B}{R}$$

R_C is the radius of curvature of any point on the involute at radius R. Inspection of Figure 2.3 reveals that R_C is also the length of string unrolled from the base circle as the base circle rolls through an angle E; therefore,

$$E \cdot R_B = R_C$$

$$R_C = \sqrt{R^2 - R_B^2}$$

$$\theta = E - \phi = \frac{R_C}{R_B} - \tan^{-1} \frac{R_C}{R_B}$$

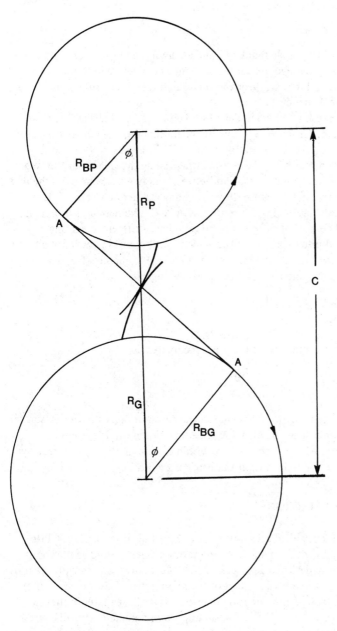

Figure 2.4 Gear teeth in mesh.

or

$$\theta = \tan \phi - \phi$$

Thus, knowing the base circle radius R_B and assuming values for ϕ, R, and θ, the polar coordinates of the involute can be plotted. The involute in terms of the pressure angle is useful in many gear tooth calculations, and a table of involutes is presented in the Appendix.

Let us now mesh two involute curves together at a center distance C, as shown in Figure 2.4. The angle ϕ is now the operating pressure angle of the gear mesh and R_P and R_G are the operating pitch radii. The subscripts P and G stand for pinion and gear, with the pinion always being the smaller of the two meshing gears. AA, the common tangent between the two base circles, is the line of action and the two involutes are shown meshing at the pitch point. It is important to understand that if the center distance C is increased or decreased, the involutes will contact at different points and have different operating pressure angles and pitch diameters. The velocity ratio, however, will not change since it is dependent only on the ratio of the two base diameters. The relationship of the base radius, pitch radius, and pressure angle is

$$\cos \phi = \frac{R_{BP}}{R_P} = \frac{R_{BG}}{R_G}$$

The center distance C in terms of the other parameters is

$$C = R_P + R_G = \frac{R_{BP} + R_{BG}}{\cos \phi}$$

The insensitivity of the involute to center distance variation is another reason for using this curve for gear teeth. Also, it can be seen that a whole system of involute gearing can be established where within certain limits any two gears will mesh with one another and transmit uniform rotary motion.

GEAR TOOTH DEFINITIONS

Figure 2.5 depicts a practical gear tooth. As shown, a portion of the involute curve bounded by the outside diameter and root diameter is used as the tooth profile. In a properly designed gear mesh the involute curve merges with the root fillet at a point below the final contact of the mating gear. This intersection of involute and root fillet, called the form diameter, is discussed later in the chapter. From Figure 2.5 we can define an important parameter, the diametral pitch, which is a measure of tooth size. The diametral pitch is the number of teeth per inch of pitch diameter:

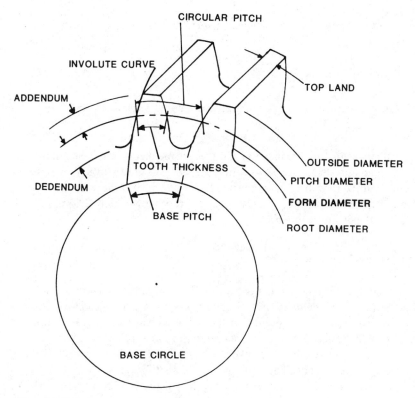

Figure 2.5 Gear tooth nomenclature.

$$DP = \frac{N}{PD}$$

where

    N = number of teeth
    PD = pitch diameter, in.

The circumference of the pitch diameter divided by the number of teeth is called the circular pitch:

$$CP = \pi \cdot \frac{PD}{N}$$

The circumference of the base diameter divided by the number of teeth is called the base pitch P_B:

$$P_B = \pi \frac{BD}{N}$$

The relationship between circular pitch and diametral pitch is

$$CP \cdot DP = \pi$$

The area between the pitch diameter and the outside diameter is called the addendum and on a standard gear tooth is 1.0/DP. The area between the outside diameter and the root diameter is the whole depth, which is the addendum plus dedendum. The whole depth of a standard gear tooth is generally 2.25 to 2.4 divided by the diametral pitch.

GEAR TOOTH GENERATION

Let us look at the generation of a gear tooth. Figure 2.6 shows a straight-sided cutting tool, such as that used in the hobbing process, generating an involute tooth. By "generating" it is meant that the tool is cutting a conjugate form. Such a straight-sided tooth is sometimes referred to as a rack. As the tool traverses and the work rotates, an involute is generated on the gear tooth flank and a trochoid in the root fillet, as shown in Figure 2.7.

Figure 2.8 is a closer look at a hob tooth. This is a hob of pressure angle ϕ and diametral pitch $\pi/(TH + TP)$. It is capable of cutting a whole family of gears with the same pressure angle and diametral pitch. Such tools are standard, as, for instance, a 20° pressure angle and 8 pitch (diametral pitch) hob, and are easily obtainable. Gears cut by this hob will be capable of meshing with one another. The distance (B + R) on the hob tooth is equal to the dedendum of the

CUTTING TOOL

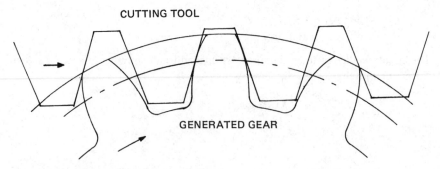

GENERATED GEAR

Figure 2.6 Generating a gear tooth.

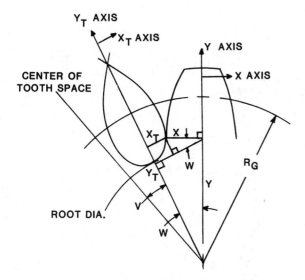

Figure 2.7 Root fillet trochoid and involute.

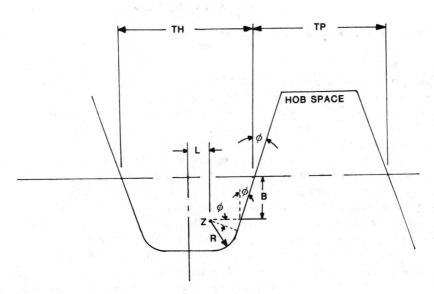

Figure 2.8 Hob geometry.

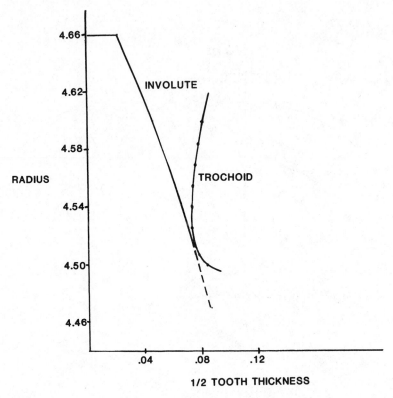

Figure 2.9   Plot of involute and trochoid.

generated gear tooth. R is the hob tip radius with its center at point Z. TP, the hob tooth space, is equal to the tooth thickness of the generated gear at the gear pitch diameter. TH is the hob tooth thickness. When the hob traverses a distance (TP + TH), the gear rotates through an angle (TP + TH)/R_G, where R_G is the gear pitch radius. (TP + TH) is the circular pitch of the gear.

The involute and trochoid can be plotted on a Cartesian coordinate system emanating at the center of the gear with the Y axis going through the center of the gear tooth, as shown in Figure 2.7. Figure 2.9 is a plot of the profile of a hobbed 20° pressure angle tooth on a Cartesian coordinate system. In the following paragraphs equations are developed to generate this plot. These equations are easily programmed and the coordinates can be plotted automatically.

First, the involute coordinates will be obtained. Previously, it was shown that if the base circle radius is known, the involute angle θ and the radius to the curve can be found for any assumed pressure angle. To find the coordinates with respect to the center of the tooth, the tooth thickness at any radius must be

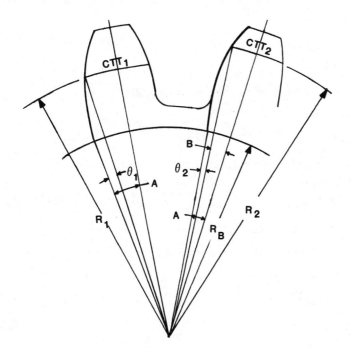

Figure 2.10 Tooth thickness calculation.

known. Let us start at the pitch diameter with a pressure angle ϕ_1, a pitch radius R_1, and a circular tooth thickness CTT_1. The involute angle is

$$\theta_1 = \tan \phi_1 - \phi_1$$

Referring to Figure 2.10, it can be seen that the angle A is

$$A = \theta_1 + \frac{\frac{1}{2}CTT_1}{R_1}$$

To find the circular tooth thickness at any other radius R_2, we use

$$B = A - \theta_2 = \theta_1 + \frac{\frac{1}{2}CTT_1}{R_1} - \theta_2$$

$$\phi_2 = \cos^{-1} \frac{R_B}{R_2}$$

$$\theta_2 = \tan \phi_2 - \phi_2$$

and

$$CTT_2 = 2R_2 \left(\frac{\frac{1}{2}CTT_1}{R_1} + \theta_1 - \theta_2 \right)$$

To find the X and Y coordinates of the involute at any radius R_2, we use

X $= R_2 \sin B$

Y $= R_2 \cos B$

Following are the steps that would be used in a computer routine to calculate the involute profile:

R_1 = pitch radius

ϕ_1 = pressure angle at pitch radius

CTT_1 = circular tooth thickness at pitch radius

R_B = base radius

θ_1 = $\tan \phi_1 - \phi_1$

ϕ_2 = 0.0

1   θ_2 = $\tan \phi_2 - \phi_2$

R_2 = $\dfrac{R_B}{\cos \phi_2}$

CTT_2 = $2R_2 \left(\dfrac{\frac{1}{2}CTT_1}{R_1} + \theta_1 - \theta_2 \right)$

B   = $\dfrac{\frac{1}{2}CTT_2}{R_2}$

X $= R_2 \sin B$

Y $= R_2 \cos B$

Write (X, Y)

If $(CTT_2 \cdot LE \cdot 0.005)$ go to 2

ϕ_2 = $\phi_2 + \dfrac{\pi}{180}$

Go to 1

2   Continue

The routine starts at the base circle, where the pressure angle is 0.0, and calculates coordinates at pressure angle intervals of $\pi/180$ rad until it reaches a point near the tip of the tooth where the thickness is less than 0.005 in.

In the hobbing process depicted in Figure 2.6, the point on the corner of the tool generates the shape of the root fillet. This shape, which is bounded by the root diameter and the involute, is called a trochoid. In Figure 2.7 the complete curve is shown as a loop on the X_T and Y_T axes. Only that portion of the curve that lies between the involute and root diameter is of interest. To calculate the trochoid coordinates, the following procedure is used:

1. Calculate the trochoid generated by the center of the hob tip radius, point Z on Figure 2.8.
2. Find the normal to this trochoid at any point in order to add the radius R. This step would not be necessary if the hob had a sharp corner since the corner point would cut a trochoid, but practical hobs have rounded tips.
3. The trochoid coordinates calculated will be with respect to the center of the trochoid (Figure 2.7, X_T and Y_T axes). To find the coordinates on the desired X-Y system through the center of the tooth they must be shifted through the angle W (Figure 2.7).

When the hob traverses a distance (TH + TP) (Figure 2.8), the gear rotates through an angle $(TH + TP)/R_G$; therefore, the angle (W + V) between the center of the gear tooth and the center of the tooth space is

$$W + V = \frac{\frac{1}{2}(TH + TP)}{R_G}$$

where R_G is the gear pitch radius in inches. Angle V in Figure 2.7 can be calculated as follows:

$$V = \frac{L}{R_G}$$

where

  L = distance between the center of the hob tooth and point Z on Figure 2.8, in.

$$L = \frac{TH}{2} - B \tan \phi - \frac{R}{\cos \phi}$$

and

$$W = \frac{\frac{1}{2}(TH + TP) - L}{R_G}$$

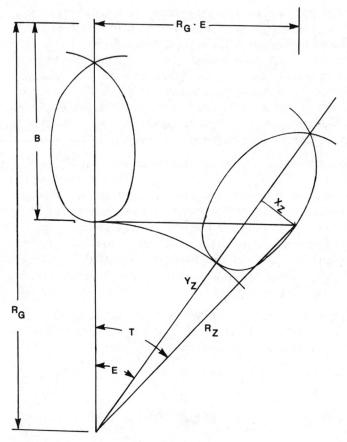

Figure 2.11 Trochoid generated by point Z.

Figure 2.11 shows the trochoid generated by point Z at its starting point and after the hob has moved a distance $R_G \cdot E$ and the gear has rotated through an angle E. The coordinates are:

$$X_Z = R_Z \sin (T - E)$$

$$= R_Z (\sin T \cos E - \cos T \sin E)$$

$$Y_Z = R_Z \cos (T - E)$$

$$= R_Z (\cos T \cos E + \sin T \sin E)$$

where

$$\cos T = \frac{R_G - B}{R_Z}$$

$$\sin T = \frac{R_G \cdot E}{R_Z}$$

Therefore,

$$X_Z = (R_G \cdot E) \cos E - (R_G - B) \sin E \qquad (2.1)$$
$$Y_Z = (R_G - B) \cos E + (R_G \cdot E) \sin E$$

Figure 2.12 shows how to calculate the actual trochoid coordinates, adding the hob tip radius R to the trochoid generated by point Z:

$$X_T = X_Z + R \cos A \qquad (2.2)$$
$$Y_T = Y_Z - R \sin A$$

A is the angle formed by a line normal to the trochoid generated by point Z and the Y_T axis and can be found as follows:

$$\tan A = \frac{dX_Z}{dY_Z}$$

To find dX_Z/dY_Z:

$$\frac{dX_Z}{dE} = -(R_G \cdot E) \sin E + R_G \cos E - R_G \cos E + B \cos E$$

$$\frac{dY_Z}{dE} = -R_G \sin E + B \sin E + (R_G \cdot E) \cos E + R_G \sin E$$

and dX_Z/dY_Z is

$$\frac{dX_Z}{dY_Z} = \frac{-(R_G \cdot E) \sin E + B \cos E}{B \sin E + (R_G \cdot E) \cos E}$$

Finally, to obtain the trochoid coordinates with respect to the system through the gear tooth center, refer to Figure 2.7:

$$\sin W = \frac{X_T + X \cos W}{Y}$$

$$\cos W = \frac{Y_T - X \sin W}{Y}$$

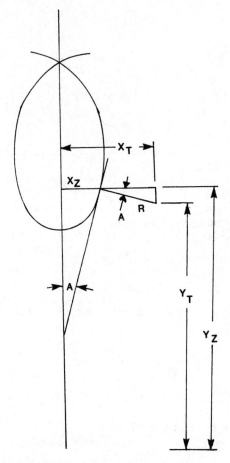

Figure 2.12 Trochoid coordinates.

$$\frac{X_T + X \cos W}{\sin W} = \frac{Y_T - X \sin W}{\cos W}$$

$$X_T \cos W + X \cos^2 W = Y_T \sin W - X \sin^2 W \qquad (2.3)$$

$$X = Y_T \sin W - X_T \cos W$$

$$Y = Y_T \cos W + X_T \sin W$$

Let us review the procedure for plotting the trochoid coordinates:

1. Choose angles E at random and calculate the trochoid coordinates of point Z using Eqs. (2.1).

2. Calculate the angle A at each point and define the actual trochoid coordinates using Eqs. (2.2).
3. Convert to the Cartesian coordinate system through the tooth center using Eqs. (2.3).

Following is a computer routine to carry out the process described above, starting with angle E equal to 0 rad (where angle A is 90° or $\pi/2$ rad) and ending with E equal to $25\pi/180$ rad.

$$\text{Counter} = 25 \cdot \frac{\pi}{180}$$

$$L = \tfrac{1}{2}\text{TH} - B \tan \phi - \frac{R}{\cos \phi}$$

$$W = \frac{(\text{TH} + \text{TP})/2 - L}{R_G}$$

$$E = 0.0$$

1    $$\frac{dX_Z}{dE} = -(R_G \cdot E) \sin E + B \cos E$$

$$\frac{dY_Z}{dE} = B \sin E + (R_G \cdot E) \cos E$$

$$X_Z = (R_G \cdot E) \cos E - (R_G - B) \sin E$$

$$Y_Z = (R_G - B) \cos E + (R_G \cdot E) \sin E$$

If $(E \cdot LE \cdot 0.0)$ go to 2

$$A = \tan^{-1} \frac{dX_Z}{dY_Z}$$

Go to 3

2    $$A = \frac{\pi}{2}$$

3    $$X_T = X_Z + R \cos A$$

$$Y_T = Y_Z - R \sin A$$

$$X = Y_T \sin W - X_T \cos W$$

$$Y = Y_T \cos W + X_T \sin W$$

If $(E \cdot GE \cdot \text{counter})$, go to 4

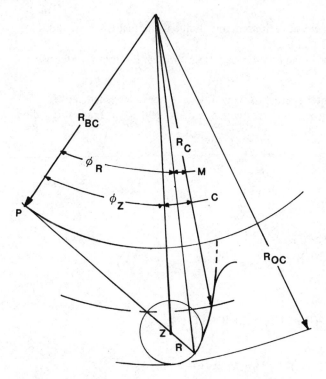

Figure 2.13 Shaper cutter tooth.

Write X, Y

$$E = E + \frac{0.25\,\pi}{180}$$

Go to 1

4   Continue

The previous discussion was concerned with a straight-sided cutter such as a hob. Another type of cutting tool is in the form of a gear tooth and is called a shaper cutter. The shaper cutter meshes with the work and generates a mating gear. Figure 2.13 shows a shaper cutter tooth with the profile in the form of an involute and with a rounded tip edge of radius R.

R_C = cutter pitch radius, in.

R_{BC} = cutter base circle radius, in.

R   = round edge radius, in.

R_{OC} = cutter outside radius, in.

Z = center of round edge

As the cutter rolls through angle C the gear it is cutting will roll through an angle $C \cdot R_C/R_G$, where R_G is the gear pitch radius. Referring back to Figure 2.7, the angle W between the center of the trochoid and the center of the gear tooth is

$$W = \frac{\frac{1}{2}CTT_G}{R_G} + \left(\frac{C \cdot R_C}{R_G}\right)$$

where CTT_G is the circular tooth thickness of the gear at the pitch diameter, in inches. From Figure 2.13 angle C is found as follows:

$$C = \phi_R + M - \phi_Z$$

$$M = Inv\, \phi_R - Inv\, \phi$$

$$\phi_Z = \cos^{-1} \frac{R_{BC}}{R_{OC} - R}$$

$$PZ = R_{BC} \tan \phi_Z$$

$$\phi_R = \tan^{-1} \frac{PZ + R}{R_{BC}}$$

where ϕ is the pressure angle at the pitch diameter.

Figure 2.14 shows how the coordinates of the trochoid of point Z, the center of the shaper cutter round edge radius, are arrived at:

$$X_Z = R_Z \sin (T - E_G)$$

$$Y_Z = R_Z \cos (T - E_G)$$

$$\sin T = (R_{OC} - R) \frac{\sin E_C}{R_Z}$$

$$\cos T = \frac{R_Z^2 + (R_C + R_G)^2 - (R_{OC} - R)^2}{2(R_Z)(R_C + R_G)}$$

$$R_Z^2 = (R_{OC} - R)^2 + (R_C + R_G)^2 - 2(R_{OC} - R)(R_C + R_G) \cos E_C$$

Therefore,

$$X_Z = (R_{OC} - R) \sin (E_C + E_G) - (R_C + R_G) \sin E_G$$

$$Y_Z = (R_C + R_G) \cos E_G - (R_{OC} - R) \cos (E_C + E_G)$$

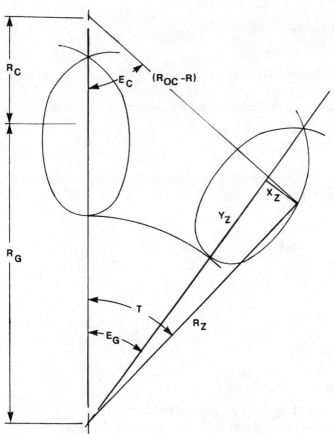

Figure 2.14 Shaper cutter trochoid coordinates.

To find the trochoid coordinates with respect to the center of the gear tooth, the same procedure is followed as was used for the hob-type cutter:

1. From Figure 2.12 find the X_T and Y_T coordinates differentiating the foregoing equations to calculate tan A using Eqs. (2.2).
2. From Figure 2.7, shift to the X, Y coordinate system by using Eqs. (2.3).

GEAR TEETH IN ACTION

Figure 2.15 shows a gear mesh with the driving pinion tooth on the left just coming into mesh at point T and the two teeth on the right meshing at point S.

Notice that contact starts at point T where the outside diameter of the gear crosses the line of action and ends where the outside diameter of the pinion crosses the line of action, point R.

Z is the length of the line of action. In other words, a tooth will be in contact from point T to point R. P_B is the base pitch, the distance from one involute to the next along a radius of curvature. It was shown earlier that

$$P_B = \pi \frac{BD}{N}$$

where

BD = base diameter, in.
N = number of teeth

Point T, where contact initiates, is called the lowest point of contact on the pinion tooth and also the highest point of contact on the gear tooth. Similarly, point R is the highest point of contact on the pinion tooth and the lowest point of contact on the gear tooth. Point S is the highest point of single tooth contact on the pinion and the lowest point of single tooth contact on the gear. In other words, if one imagines the gears in Figure 2.15 to begin rotating, just prior to

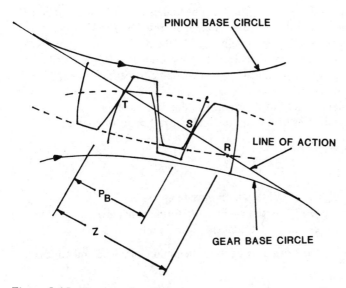

Figure 2.15 Gear tooth action.

meshing at point T a single pair of teeth was carrying the load. As the gears continue turning and the pinion tooth on the right moves from point S to R, two pairs of teeth are carrying the load and after point R a single pair again carries the load until the next two teeth mesh at point T.

Thus it can be seen that for some period of time one tooth mesh carries the load and for another period of time two tooth meshes share the load. A measure of the percentage of time two meshes share the load is the profile contact ratio M_p. For instance, a profile contact ratio of 1.0 would mean that one tooth is in contact 100% of the time. A contact ratio of 1.6 means that two pairs of teeth are in contact 60% of the time and one pair carries the load 40% of the time. Contact ratios for conventional gearing are generally in the range 1.4 to 1.6.

Let us now derive the profile contact ratio in terms of parameters easily obtainable:

$$M_P = \frac{E_{TR}}{360°/N}$$

where

E_{TR} = degrees of roll to traverse the length of the line of action from point T to R

N = number of teeth

This equation may not be obvious, but it can be understood if it is remembered that from one tooth to another the base circle must roll 360°/N. Thus if the base circle rolls 360°/N while going from point T to R, the profile contact ratio is 1.0. If the base circle rolls more than 360°/N going from T to R, the profile contact ratio is greater than 1, indicating that for some percentage of time two pairs of teeth are in contact.

E_{TR}, the total degrees of roll, is equal to the degrees of roll to the pinion tooth tip minus the degrees of roll to the pinion lowest point of contact:

$$E_{TR} = E_{ODP} - E_{TIFP}$$

where

E_{ODP} = degrees of roll to pinion outside diameter

E_{TIFP} = degrees of roll to true involute form diameter on the pinion (lowest point of contact) (point T on Figure 2.15)

Figure 2.16 shows how the degrees of roll to the pinion outside diameter is calculated:

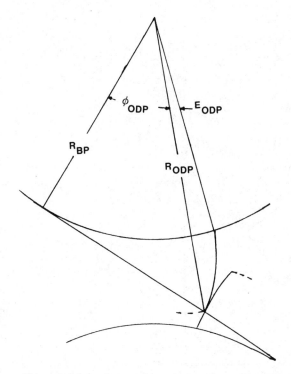

Figure 2.16 Degrees of roll to pinion outside diameter.

$$E_{ODP} = \tan \phi_{ODP}\left(\frac{180}{\pi}\right) = \frac{\sqrt{R_{ODP}^2 - R_{BP}^2}}{R_{BP}}\left(\frac{180}{\pi}\right)$$

Figure 2.17 shows how the degrees of roll to the pinion form diameter is calculated:

$$E_{TIFP} = \tan \phi_{TIFP}\left(\frac{180}{\pi}\right) = \frac{\sqrt{R_{TIFP}^2 - R_{BP}^2}}{R_{BP}}\left(\frac{180}{\pi}\right)$$

It is more convenient to express E_{TIFP} in terms of the gear outside radius and base radius:

$$XX = C \sin \phi_{PD}$$

$$\sqrt{R_{TIFP}^2 - R_{BP}^2} = C \sin \phi_{PD} - \sqrt{R_{ODG}^2 - R_{BG}^2}$$

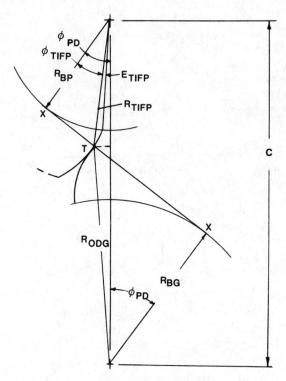

Figure 2.17 Degrees of roll to pinion true involute form diameter.

Therefore,

$$E_{TIFP} = \frac{C \sin \phi_{PD} - \sqrt{R_{ODG}^2 - R_{BG}^2}}{R_{BP}} \left(\frac{180}{\pi} \right)$$

and

$$M_P = \frac{\sqrt{R_{ODP}^2 - R_{BP}^2} - C \sin \phi_{PD} + \sqrt{R_{ODG}^2 - R_{BG}^2}}{R_{BP}} \frac{180N}{\pi \cdot 360}$$

where

$$N = \text{number of pinion teeth}$$
$$R_{BP} = \tfrac{1}{2}PD_P \cos \phi_{PD}$$
$$\frac{N}{PD_P \cdot \pi} = \frac{1}{CP}$$

Therefore,

$$M_P = \frac{\sqrt{R_{ODP}^2 - R_{BP}^2} - C \sin \phi_{PD} + \sqrt{R_{ODG}^2 - R_{BG}^2}}{CP \cos \phi_{PD}}$$

Another way of expressing the profile contact ratio is

$$CP \cos \phi_{PD} = P_B$$

because

$$\cos \phi_{PD} = \frac{BD}{PD}$$

and

$$CP = \pi \frac{PD}{N}$$

Therefore,

$$CP \cos \phi_{PD} = \pi \frac{BD}{N} = P_B$$

and

$$\sqrt{R_{ODP}^2 - R_{BP}^2} - C \sin \phi_{PD} + \sqrt{R_{ODG}^2 - R_{BG}^2} = Z$$

Therefore,

$$M_P = \frac{Z}{P_B}$$

To calculate the degrees of roll to the highest point of single tooth contact on the pinion, consider Figure 2.18. The distance XS is the sum of XT from Figure 2.18 and TS from Figure 2.15. The distance XT can be calculated using Figure 2.17:

$$XT = \sqrt{R_{TIFP}^2 - R_{BP}^2}$$

From Figure 2.15 it is seen that the distance TS is a base pitch; therefore,

$$XS = \sqrt{R_{TIFP}^2 - R_{BP}^2} + P_B$$

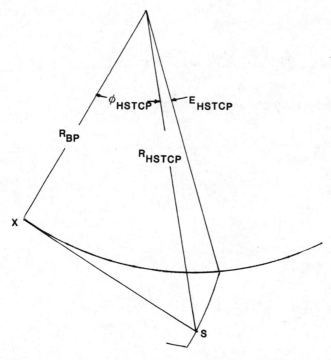

Figure 2.18 Degrees of roll to pinion highest single tooth contact diameter.

$$E_{HSTCP} = \tan \phi_{HSTCP}\left(\frac{180}{\pi}\right) = \frac{XS}{R_{BP}}\left(\frac{180}{\pi}\right)$$

$$= \frac{\sqrt{R_{TIFP}^2 - R_{BP}^2} + P_B}{R_{BP}}\left(\frac{180}{\pi}\right)$$

or

$$E_{HSTCP} = E_{TIFP} + \frac{2\pi}{N}$$

The involute curve changes very rapidly near the base diameter and more slowly at sections farther away from the base circle. This is illustrated in Figure 2.19, which shows the difference in length along the involute curve for equal increments taken on the base circle. The distance XY is far less than YZ. Since the

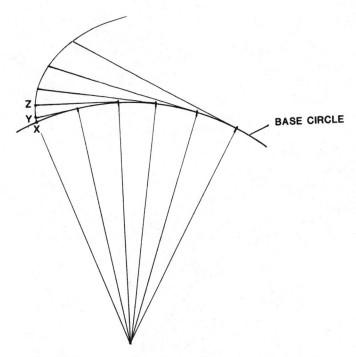

Figure 2.19 Involute curve properties.

involute is so sensitive near the base circle, the lowest point of contact on a gear tooth should be located well away from the base circle. As a rule of thumb the lowest point of contact on a gear tooth should be at least $9°$ of roll.

ROLLING AND SLIDING VELOCITIES

When involute gear teeth mesh, the action is not pure rolling as it would be when two friction disks are in contact, but a combination of rolling and sliding. Figure 2.20 shows a gear mesh with two base circles of equal size and the teeth meshing at the pitch point. Radii of curvature are drawn to the involutes from equal angular intervals on the base circle. It can be seen that arc XY on gear 2 will mesh with arc AB on gear 1 and that AB is longer than XY; therefore, the two profiles must slide past one another to make up the difference in length. The sliding velocity, which is usually expressed in feet per minute, at any point is calculated as follows:

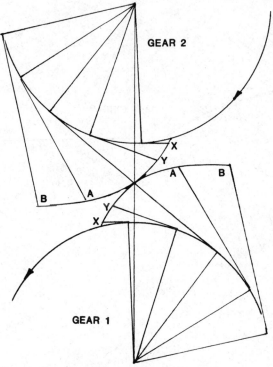

Figure 2.20   Relative sliding of gear teeth.

$$V_S = \frac{W_1 R_{C1} - W_2 R_{C2}}{12}$$

where

V_S = sliding velocity, fpm
W_1 = angular velocity of gear 1, rad/min
W_2 = angular velocity of gear 2, rad/min
R_{C1} = radius of curvature of gear 1, in.
R_{C2} = radius of curvature of gear 2, in.

From Figure 2.20 it can be seen that when point A on gear 1 and point Y on gear 2 mesh, R_{C1} will be larger than R_{C2} and since $W_1 = W_2$, V_S will be a positive number. As the meshing point nears the pitch point the difference in the radii of curvature lessens until at the pitch point the radii of curvature are equal and V_S is 0. When point A on gear 2 meshes with point Y on gear 1, R_{C1} will be smaller than R_{C2} and V_S will be negative. The significance of this is that as the mesh goes through the pitch point the direction of sliding changes. There is always pure rolling at the pitch point. If the base circles were of unequal size at the pitch point, $W_1 R_{C1}$ would still equal $W_2 R_{C2}$ since

$$\frac{W_1}{W_2} = \frac{R_2}{R_1} = \frac{R_{C1}/\sin\phi}{R_{C2}/\sin\phi}$$

where

    R_1 = pitch radius of gear 1, in.
    R_2 = pitch radius of gear 2, in.
    ϕ = pressure angle, deg

Sliding velocity is significant in that it affects the amount of heat generated in the gear mesh. Also, the fact that gear teeth undergo sliding as well as rolling must be appreciated.

Another significant velocity term is the sum or entraining velocity:

$$V_E = W_1 R_{C1} + W_2 R_{C2}$$

where V_E is the sum velocity in ips. The sum or entraining velocity is a measure of how quickly oil is being dragged into the conjunction between the two gear members.

The parameter generally used when expressing the speed of a gearset is the pitch line velocity:

$$V_T = \frac{W_P R_P (60)}{12} = \frac{W_G R_G (60)}{12}$$

$$= \frac{\pi D_P n_P}{12} = \frac{\pi D_G n_G}{12}$$

where

    V_T = pitch line velocity, fpm
    W_P = pinion angular velocity, rad/sec
    W_G = gear angular velocity, rad/sec
    R_P = pinion pitch radius, in.
    R_G = gear pitch radius, in.
    D_P = pinion pitch diameter, in.
    D_G = gear pitch diameter, in.
    n_P = pinion rpm
    n_G = gear rpm

The pitch line velocity is a measure of the tangential or peripheral velocity of a gearset and a better indication of speed than the rpm. For instance, a 1-in. pitch

diameter gear operating at 10,000 rpm has the same pitch line velocity as a
10-in. pitch diameter gear operating at 1000 rpm. Two meshing gears always
have the same pitch line velocity.

American Gear Manufacturers Association (AGMA) Standards for enclosed
drives consider units with pitch line velocities of 5000 fpm or more high speed.
Gear units have been operated at pitch line velocities up to 50,000 fpm minute
but applications over approximately 20,000 fpm require extremely careful
analysis concerning lubrication, cooling, and centrifugal effects.

HELICAL GEARS

Figure 1.13 illustrates the difference between spur and helical gears. The tooth
contact on spur gears is a straight line across the tooth and at any time either
one or two teeth are in contact. The helical gear contact, because the teeth are at
an angle to the axis of rotation, is a series of oblique lines with several teeth in
contact simultaneously and the total length of contact varies as the teeth go
through the mesh.

To understand the nature of the helical tooth, consider a base cylinder
with a series of strings wrapped around it as shown in Figure 2.21. The start of
each string is offset such that a line joining the string starts is at an angle ψ_B to
the axis of rotation of the cylinder. The ends of each string when held taut and
unwrapped from the base cylinder will define involutes and the surface defined
by the string ends will be a helical involute gear tooth.

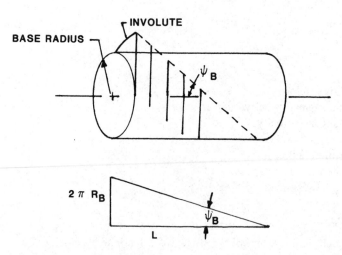

Figure 2.21 Helical gear base cylinder.

In one rotation of the base cylinder an axial length L of strings will be unwrapped. L is defined as the lead

$$L = \frac{2\pi R_B}{\tan \psi_B}$$

where

L = lead, in.
R_B = base radius, in.
ψ_B = base helix angle, deg

The helix angle along the tooth profile varies with the radius but the lead is a constant. Once the base radius and base helix angle are defined, the lead can be calculated and the helix angle at any radius R is known:

$$\tan \psi_R = \frac{2\pi R}{L}$$

where ψ_R is the helix angle at radius R, in degrees.

Transverse and Normal Planes

Figure 2.22 shows the relationship between the transverse and normal planes of a helical gear. The transverse plane ABCD is the plane of rotation, while the normal plane ABE is at right angles to the tooth. The normal and transverse planes are displaced from each other through the helix angle ψ. Following are the relationships between the normal and transverse pressure angles at any radius on the tooth illustrated at point B on Figure 2.22.

$$\tan \phi_N = \frac{AB}{AE}$$

$$\tan \phi_T = \frac{AB}{AD}$$

$$\cos \psi = \frac{AD}{AE}$$

$$AD \tan \phi_T = AE \tan \phi_N$$

Therefore,

$$\cos \psi \tan \phi_T = \tan \phi_N$$

Figure 2.23 shows a helical gear rotating about the axis XX. The teeth are inclined with relation to the axis of rotation the helix angle ψ. Usually, the helix angle at the operating pitch diameter is referred to. When the inclination of the

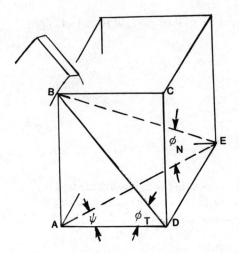

Figure 2.22 Normal and transverse planes.

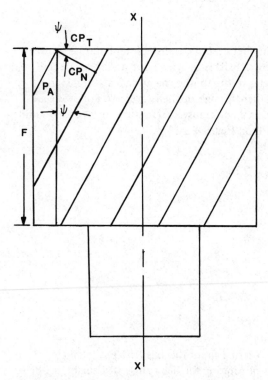

Figure 2.23 Normal and transverse pitches.

teeth is off to the right as shown in Figure 2.23, the gear helix is designated as right hand. When two external helical gears mesh, one must be right hand and the other left hand. When an external gear meshes with an internal gear they will both have the same hand of helix.

The circular pitch in the transverse plane CP_T has the following relationship with the circular pitch CP_N in the normal plane:

$$\cos \psi = \frac{CP_N}{CP_T}$$

The circular tooth thicknesses have the same relationship:

$$\cos \psi = \frac{CTTN}{CTTT}$$

where

$$\begin{aligned} CTTN &= \text{normal circular tooth thickness, in.} \\ CTTT &= \text{transverse circular tooth thickness, in.} \end{aligned}$$

The normal and transverse diametral pitches have the relationship

$$DP_N \cos \psi = DP_T$$

The distance along the tooth axis from one tooth to another is called the axial pitch P_A, as shown in Figure 2.23. The ratio of face width F to axial pitch is called the face contact ratio or the helical overlap and is a measure analogous to the profile contact ratio for spur gears. The face contact ratio, designated as M_F, is

$$M_F = \frac{F}{P_A}$$

$$P_A = \frac{CP_N}{\sin \psi} = CP_T \frac{\cos \psi}{\sin \psi} = \frac{\pi}{DP_T \tan \psi}$$

Therefore,

$$M_F = F(DP_T) \frac{\tan \psi}{\pi}$$

The total contact of a helical gear mesh is therefore some combination of the profile and face contact ratios. Sometimes the sum of the two is called the total contact ratio and used as a measure of how much contact is achieved in a tooth mesh.

The actual total length of contact at any instant in a helical mesh is the sum of the length of the oblique lines of contact on each tooth in mesh and varies as the teeth go through mesh. A method of calculating the minimum and maximum length of the lines of contact was derived by E. J. Wellauer and presented to the Industrial Mathematics Society. The paper was entitled "The Nature of the Helical Gear Oblique Contact Line," and a small portion based on the original article is given below.

K_a and n_a are the whole number and fractional portion, respectively, of the face contact ratio. For example, if $M_F = F/P_A = 4.85$, then $K_a = 4.0$ and $n_a = 0.85$. K_r and n_r are the whole number and fractional portion, respectively, of the profile contact ratio. For example, if $M_P = Z/P_B = 1.32$, then $K_r = 1.0$ and $n_r = 0.32$.

If $(1 - n_r)/n_a \geq 1$, then

$$L_{min} = \frac{(Z \cdot F/P_B) - n_r n_a P_A}{\cos \psi_B}$$

where

L_{min} = minimum total length of the oblique lines of contact, in.
Z = length of the line of action, in. (Figure 2.15)
P_B = base pitch, in.
P_A = axial pitch, in.
ψ_B = base helix angle, deg

If $(1 - n_r)/n_a < 1$, then

$$L_{min} = \frac{(Z \cdot F/P_B) - (1 - n_a)(1 - n_r)P_A}{\cos \psi_B}$$

An approximation used for calculating L_{min} in several AGMA Standards is

$$L_{min} = \frac{0.95 \,(Z)F}{P_B \cos \psi_B}$$

To calculate the maximum total length of the lines of contact:
If $n_r \leq n_a$, then

$$L_{max} = \frac{(Z \cdot F/P_B) + n_r(1 - n_a)P_A}{\cos \psi_B}$$

If $n_r > n_a$, then

$$L_{max} = \frac{(Z \cdot F/P_B) + n_a(1 - n_r)P_A}{\cos \psi_B}$$

INTERNAL GEARS

The involute form of internal gears (sometimes called ring gears) is the same as for external gears. The difference between the two lies in the fact that internal gears contact on the concave side of the involute rather than the convex. Also, the root diameter of an internal gear tooth is the largest diameter and the tips of the teeth are at the inside diameter, which is the smallest. Figure 2.24 shows internal gear tooth geometry. At any radius R, with pressure angle ϕ and base radius R_B, all the involutometry calculations will be essentially the same as those previously shown for external gear teeth.

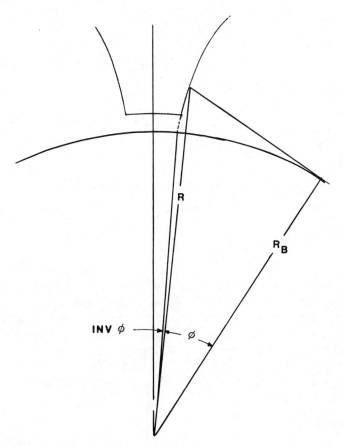

Figure 2.24 Internal gear geometry.

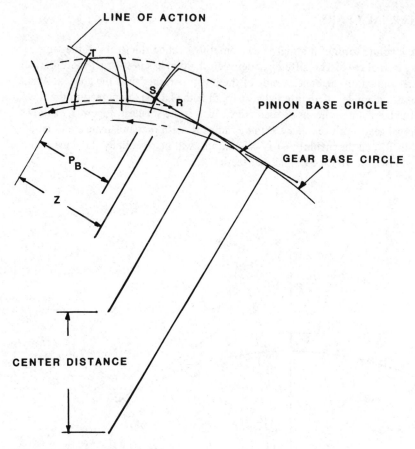

Figure 2.25 Internal gear mesh action.

To illustrate internal gear mesh calculations, let us derive the profile contact ratio for the situation shown in Figure 2.25 for an external pinion driving an internal gear. The start of contact is at point R, where the internal gear inside diameter crosses the line of action. Contact ends at point T, where the pinion outside diameter crosses the line of action. The total length of contact is the distance from R to T or Z as shown on Figure 2.25. One pair of teeth is meshing at point S on the figure and an adjacent pair at point T; therefore, the distance ST is a base pitch P_B. The profile contact ratio is

$$M_P = \frac{E_{TR}}{360°/N}$$

where

E_{TR} = degrees of roll to traverse the length of the line of action from point R to T

N = number of pinion teeth

E_{TR}, the total degrees of roll, is equal to the degrees of roll to the pinion outside diameter minus the degrees of roll to the lowest point of contact on the pinion (TIF, true involute form diameter).

$$E_{TR} = E_{ODP} - E_{TIFP}$$

where

E_{ODP} = degrees of roll to the pinion outside diameter

E_{TIFP} = degrees of roll to the pinion true involute form diameter

As shown previously (Figure 2.16),

$$E_{ODP} = \tan \phi_{ODP} = \left(\frac{180}{\pi}\right) \frac{\sqrt{R_{ODP}^2 - R_{BP}^2}}{R_{BP}} \left(\frac{180}{\pi}\right)$$

where

R_{ODP} = pinion outside radius, in.

R_{BP} = pinion base radius, in.

Figure 2.26 shows how the degrees of roll to the pinion form diameter is calculated:

$$E_{TIFP} = \tan \phi_{TIFP} = \left(\frac{180}{\pi}\right) \frac{\sqrt{R_{TIFP}^2 - R_{BP}^2}}{R_{BP}} \left(\frac{180}{\pi}\right)$$

It is more convenient to express E_{TIFP} in terms of the gear inside radius and base radius. From Figure 2.26,

$$XX = C \sin \phi_{PD}$$

where ϕ_{PD} is the pressure angle at the pitch diameter, in degrees.

$$\sqrt{R_{TIFP}^2 - R_{BP}^2} = \sqrt{R_{IDG}^2 - R_{BG}^2} - C \sin \phi_{PD}$$

where

R_{IDG} = gear inside radius, in.

R_{BG} = gear base radius, in.

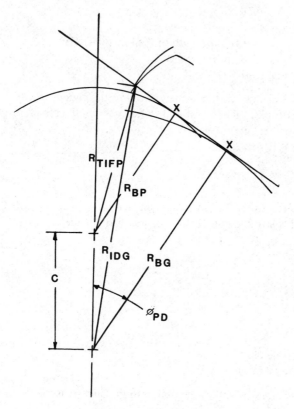

Figure 2.26   Internal gear mesh degrees of roll to pinion form diameter.

Therefore,

$$E_{TIFP} = \frac{\sqrt{R^2_{IDG} - R^2_{BG}} - C \sin \phi_{PD}}{R_{BP}} \left(\frac{180}{\pi}\right)$$

and

$$M_P = \frac{\sqrt{R^2_{ODP} - R^2_{BP}} + C \sin \phi_{PD} - \sqrt{R^2_{IDG} - R^2_{BG}}}{R_{BP}} \quad \frac{180N}{\pi \cdot 360}$$

$$R_{BP} = \tfrac{1}{2}PD_P \cos \phi_{PD}$$

$$\frac{N}{PD_P} \pi = \frac{1}{CP}$$

where CP is the circular pitch in inches. Therefore,

$$M_P = \frac{\sqrt{R_{ODP}^2 - R_{BP}^2} + C \sin \phi_{PD} - \sqrt{R_{IDG}^2 - R_{BG}^2}}{CP \cos \phi_{PD}}$$

MEASUREMENT OVER BALLS OR WIRES

This subject is presented at this point not only because it is an important measurement in the manufacture of gear teeth but because it is a good illustration of the application of involutometry in the analysis of gear tooth geometry.

When cutting or grinding a gear tooth the machine operator will check the tooth thickness to determine when sufficient stock has been removed from the flank to bring the tooth to the required size. The drawing requirement may call for a tooth thickness at a given diameter. This is a difficult measurement to make directly; therefore, quite often an indirect measurement is used. Balls or wires (sometimes called pins) of a known diameter are placed in 180° opposite tooth spaces on the gear and an accurate micrometer measurement over the balls or wires is made.

The equations for calculating measurement over balls or wires will be derived first for an external spur gear with an even number of teeth. In this case two opposite tooth spaces will be in line. The analysis will then be extended to gears with odd numbers of teeth where the opposite tooth spaces are not in line and then internal gears and helical gears will be addressed.

Referring to Figure 2.27, for a spur gear with an even number of teeth the measurement over wires (MOW) is

$$MOW = 2R_2 + 2X$$

where

R_2 = radius to the center of the wire, in.
X = wire radius, in.

We are going to calculate the MOW for a gear with a known circular tooth thickness T at a known radius R_1.

$$R_2 = \frac{R_1 \cos \phi_1}{\cos \phi_2}$$

where

ϕ_1 = pressure angle at the radius R_1, deg
ϕ_2 = pressure angle at radius R_2, deg

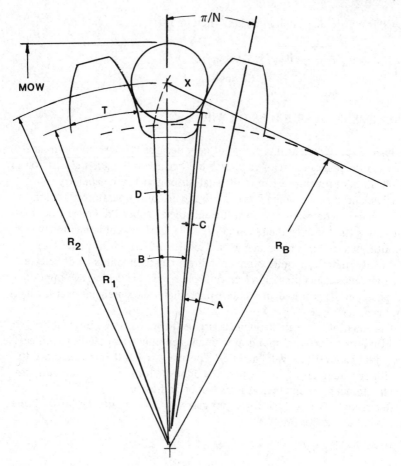

Figure 2.27 Measurement over wires of external spur gear.

The problem now is to calculate the angle ϕ_2. To accomplish this, an imaginary involute is drawn through the center of the wire as shown by the dashed profile in Figure 2.27. From this construction we can see that angle D is the involute of ϕ_2.

$$D = \mathrm{Inv}\,\phi_2 = B + C + A - \frac{\pi}{N}$$

where

       N = number of gear teeth
       C = inv ϕ_1, which can be calculated knowing ϕ_1; Inv ϕ_1 = tan ϕ_1 − ϕ_1

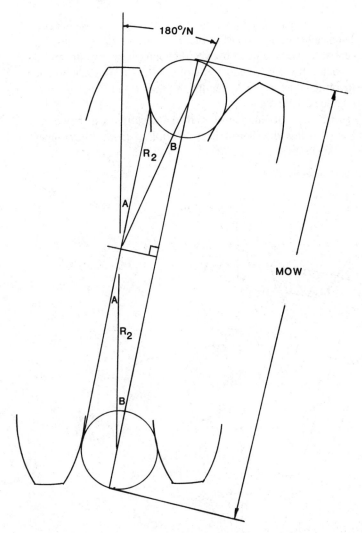

Figure 2.28 Measurement over wires for gear with odd number of teeth.

$$A = \frac{T}{2R_1}$$

$$B = \frac{X}{R_B}, \text{ where } R_B = \text{gear base radius}$$

$B = X/R_B$ because the circular distance between two involutes on the base circle is equal to the distance between normals to the involutes. In other words, the wire radius X is equal to the base pitch between the imaginary involute and the adjacent tooth involute. Knowing the involute of ϕ_2, the angle ϕ_2 can be calculated using the involute tables in the Appendix and then R_2 and the measurement over wires can be calculated.

When the gear has an odd number of teeth the situation is as shown in Figure 2.28 and the fact that the tooth spaces do not line up must be compensated for mathematically. In Figure 2.28:

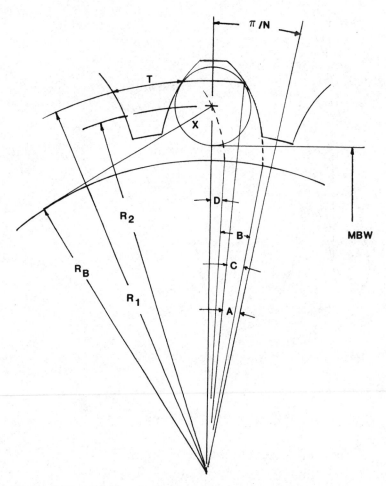

Figure 2.29 Measurement between wires of internal spur gear.

$$\angle A = \angle B = \frac{90°}{N}$$

and

$$MOW = 2\left[R_2 \cos\left(\frac{90}{N}\right) + X\right]$$

Figure 2.29 illustrates the analysis for the measurement between wires for an integral gear with an even number of teeth:

$$MBW = 2R_2 - 2X$$

$$R_2 = R_1 \frac{\cos \phi_1}{\cos \phi_2}$$

$$\text{Inv } \phi_2 = D = \frac{\pi}{N} - B - A + C$$

Again an imaginary involute is drawn through the center of the wire and the involute ϕ_2 is calculated from which the angle ϕ_2 and R_2 can be derived. For odd numbers of teeth,

$$MBW = 2\left[R_2 \cos\left(\frac{90°}{N}\right) - X\right]$$

Figure 2.30 illustrates a ball placed between two helical gear teeth. When measuring helical gears balls should be used rather than wires since the wires will not seat properly in the helices. The balls will contact the gear teeth in the normal plane, but the measurement over balls calculation must be made in the transverse plane. Figure 2.30 shows how the projection of the ball is mathematically shifted into the transverse plane and the equation for the involute ϕ_2 is

$$\text{Inv } \phi_2 = \frac{X}{R_B \cos \psi_B} + \text{Inv } \phi_1 + \frac{T}{2R_1} - \frac{\pi}{N}$$

for external gears and

$$\text{Inv } \phi_2 = - \frac{X}{R_B \cos \psi_B} + \text{Inv } \phi_1 - \frac{T}{2R_1} + \frac{\pi}{N}$$

for internal gears. It should be noted that the circular tooth thickness T is in the transverse plane. If a normal tooth thickness is given, it should be shifted to the transverse plane: $T = T_N/\cos \psi$.

To sum up, the measurement over or between balls or wires for gears with even numbers of teeth is

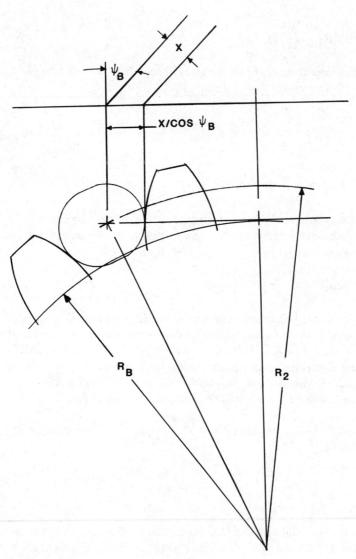

Figure 2.30 Measurement over balls of external helical gear.

$$\text{MOW} = 2(R_2 + X) \text{ for external gears}$$

$$\text{MBW} = 2(R_2 - X) \text{ for internal gears}$$

For gears with odd numbers of teeth,

$$\text{MOW} = 2\left[R_2 \cos\left(\frac{90}{N} + X\right)\right] \quad \text{for external gears}$$

$$\text{MBW} = 2\left[R_2 \cos\left(\frac{90}{N} - X\right)\right] \quad \text{for internal gears}$$

For all gears,

$$R_2 = R_1 \frac{\cos \phi_1}{\cos \phi_2}$$

To find $\cos \phi_2$,

$$\text{Inv } \phi_2 = \frac{X}{R_B \cos \psi_B} + \text{Inv } \phi_1 + \frac{T}{2R_1} - \frac{\pi}{N}$$

for external gears and

$$\text{Inv } \phi_2 = - \frac{X}{R_B \cos \psi_B} + \text{Inv } \phi_1 - \frac{T}{2R_1} + \frac{\pi}{N}$$

for internal gears. The cosine ϕ_2 is found from Inv ϕ_2 using involute tables. For spur gears, $\cos \psi_B = 0.0$.

When choosing ball or wire size a good estimate for the diameter is

$$D_{BALL} = 2X = \frac{1.728}{DP_N}$$

where

X = ball or wire radius, in.
DP_N = normal diametral pitch

Let us work through an example to illustrate the calculation for measurement over balls. Assume an external helical gear with the following dimensions:

Number of teeth	38
Normal diametral pitch	15.868103
Normal pressure angle	20.0°
Helix angle at pitch diameter	18.0°
Normal circular tooth thickness at pitch diameter, in.	0.0952

The transverse diametral pitch is

$$DP_T = 15.868103(\cos 18.0) = 15.091463$$

The transverse pressure angle is

$$\phi_T = \tan^{-1} \frac{\tan 20.0}{\cos 18.0} = 20.941896$$

The pitch diameter is

$$PD = \frac{38}{15.091463} = 2.517980$$

The base diameter is

$$BD = 2.517980(\cos 20.941896) = 2.351651$$

The lead is

$$L = \frac{\pi(2.517980)}{\tan 18} = 24.345915$$

The base helix angle is

$$\psi_B = \tan^{-1}\left(\frac{\pi(2.351651)}{24.345915}\right) = 16.880766°$$

The involute of the transverse pressure angle is

$$\text{Inv } \phi = \tan 20.941896 - 20.941896\left(\frac{\pi}{180}\right) = 0.017196$$

The transverse circular tooth thickness is

$$CTTT = \frac{0.0952}{\cos 18.0} = 0.100099$$

For a ball diameter of 0.125,

$$\text{Inv } \phi_2 = \frac{0.125/2}{(2.351651/2)\ \cos\ 16.880766} + 0.017196 + \frac{0.100099}{2.517980}$$

$$- \frac{\pi}{38} = 0.029824$$

Using the involute tables (Appendix) yields

$$\phi_2 = 24.9599°$$

and

$$R_2 = \frac{(2.517980/2) \cos 20.941896}{\cos 24.9599} = \frac{2.593914}{2}$$

The measurement over balls for 0.125-in.-diameter balls is

$$MOB = 2 \left(\frac{2.593914}{2} + \frac{0.125}{2} \right) = 2.7189$$

Measurement of Tooth Thickness by Calipers

Tooth thickness can be checked by measuring across several teeth with vernier calipers as shown on Figure 2.31. The calipers contact the teeth at points X and the line XX is tangent to the base circle. The arc AB along the base circle is equal to the length XX:

$$M = R_B \left(\frac{T}{R} + \frac{2\pi S}{N} + 2 \operatorname{Inv} \phi \right)$$

where

M = caliper measurement, in.
R_B = base circle radius, in.
T = given transverse tooth thickness at a radius R, in.
R = given radius, in.
ϕ = given transverse pressure angle at radius R, in.
S = number of tooth spaces between the contacting profiles
N = number of teeth in the gear

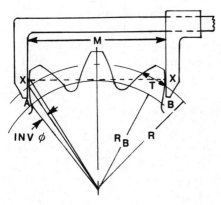

Figure 2.31 Tooth thickness measurement by vernier calipers.

For helical gears,

$$M = R_B \cos \psi_B \left(\frac{T}{R} + \frac{2\pi S}{N} + 2 \, \text{Inv} \, \phi \right)$$

where ψ_B is the base helix angle in degrees.

Center Distance and Tooth Thickness

The center distance of a pair of meshing spur or helical gears is established by the location of the centers of the bearing bores locating the gear shafts. For the gears to mesh properly at a given center distance, the tooth thicknesses must be chosen such that the teeth will not bind under all operating conditions. The following variations in center distance must be taken into account when designing a gear set:

1. The center distance will vary due to tolerances in the bearing housings.
2. Clearance in the supporting bearings will affect the operating center distance.
3. Temperature variations during operation will change the operating center distance. At a minimum, expansion of the gear teeth must be taken into account. If dissimilar materials with varying coefficients of expansion are used in the gearbox, their thermal growths must be analyzed.

   In order to accommodate all these variables, backlash is designed into the gear mesh. Backlash can be defined as the circular pitch minus the sum of the circular tooth thicknesses:

$$BL = CP - (T_P + T_G)$$

In most cases excessive backlash will not be harmful and is much more desirable than too little backlash, which can result in tight meshing and binding of the gears. In very high speed helical gearing it is important to have sufficient backlash to allow the air-oil mixture being pumped between the teeth to exit the mesh without becoming excessively churned and heated. The amount of backlash designed into a gear mesh will vary with the diametral pitch of the teeth. Following is a table of suggested backlash versus pitch.

Diametral pitch	Nominal backlash, (in.)
6	0.015
8 and 10	0.010
12 and 14	0.009
16	0.008
18	0.007
20	0.006

Excessive backlash may be detrimental if the transmitted load varies to the extent that the tooth can become unloaded and contact on the normally unloaded face. In this case the more backlash there is in the mesh, the greater freedom the teeth will have to rattle around and the greater dynamic load generated. There are also cases where gears are used as positioning devices and backlash is detrimental. In such designs special techniques such as adjustable center distance are used to achieve low backlash.

Center distance can be expressed mathematically in various ways:

$$C = \frac{PD_P + PD_G}{2} \quad \text{for external gears}$$

$$C = \frac{PD_G - PD_P}{2} \quad \text{for internal gears}$$

where

C = center distance, in.
PD_P = pinion pitch diameter, in.
PD_G = gear pitch diameter, in.

Since $PD_P/N_P = PD_G/N_G = DP_T$,

$$C = \frac{N_P + N_G}{2DP_T} \quad \text{for external gears}$$

$$C = \frac{N_G - N_P}{2DP_T} \quad \text{for internal gears}$$

where

N_P = number of pinion teeth
N_G = number of gear teeth
DP_T = transverse diametral pitch

If R is the gear ratio PD_G/PD_P,

$$C = \frac{PD_P(1 + R)}{2} \quad \text{for external gears}$$

$$C = \frac{PD_P(R - 1)}{2} \quad \text{for internal gears}$$

Let us look at a so-called standard spur gearset, where a standard diametral pitch cutter is used:

$$DP_T = 10.0$$

$$\phi_{T1} = 20°$$

$$N_P = 20$$

$$N_G = 30$$

$$C_1 = \frac{20 + 30}{2(10)} = 2.5$$

If the backlash is 0.010,

$$T_{P1} = T_{G1} = \frac{CP}{2} - \frac{BL}{2} = \frac{\pi}{(10)2} - \frac{0.010}{2} = 0.152080$$

where

T_{P1} = pinion transverse circular tooth thickness at the pitch radius R_{P1}
T_{G1} = gear transverse circular tooth thickness at the pitch radius R_{G1}

Let us calculate at what center distance C_2 the mesh will have zero backlash. At this point of tight mesh or binding,

$$CP = T_{P2} + T_{G2} = \frac{2\pi R_{P2}}{N_P} = \frac{2\pi R_{G2}}{N_G}$$

where

CP = transverse circular pitch, in.
T_{P2} = pinion transverse circular tooth thickness at the tight mesh pitch radius R_{P2}, in.
T_{G2} = gear transverse circular tooth thickness at the tight mesh pitch radius R_{G2}, in.

As shown previously:

$$T_{P2} = 2R_{P2}\left(\frac{T_{P1}}{2R_{P1}} + \text{Inv } \phi_1 - \text{Inv } \phi_2\right)$$

$$T_{G2} = 2R_{G2}\left(\frac{T_{G1}}{2R_{G1}} + \text{Inv } \phi_1 - \text{Inv } \phi_2\right)$$

Also,

$$\frac{N_G}{N_P} = \frac{R_{G1}}{R_{P1}} = \frac{R_{G2}}{R_{P2}}$$

Combining the four equations above, we have

$$\text{Inv } \phi_2 \;=\; \frac{N_P(T_{P1} + T_{G1}) \;-\; 2R_{P1}\pi}{2R_{P1}(N_P + N_G)} \;+\; \text{Inv } \phi_1$$

$$\frac{\cos \phi_2}{\cos \phi_1} \;=\; \frac{R_{P1}}{R_{P2}} \;=\; \frac{R_{G1}}{R_{G2}}$$

and the tight mesh center distance $C_2 = C_1 \cos \phi_1 / \cos \phi_2$. For the example above,

$$\text{Inv } \phi_2 \;=\; 0.012904$$

$$\phi_2 \;=\; 19.0910°$$

$$C_2 \;=\; 2.4860$$

For internal gears the equation for T_{G2} is

$$T_{G2} \;=\; 2R_{G2}\left(\frac{T_{G1}}{2R_{G1}} \;-\; \text{Inv } \phi_1 \;+\; \text{Inv } \phi_2 \right)$$

and

$$\text{Inv } \phi_2 \;=\; \frac{N_P(T_{P1} + T_{G1}) \;-\; 2R_{P1}\pi}{2R_{P1}(N_P - N_G)} \;+\; \text{Inv } \phi_1$$

In this case T_{G1} and T_{G2} are circular tooth thicknesses of the internal gear.

The situation is somewhat more complicated for helical gears since standard diametral pitch, pressure angle, and tooth thickness are defined in the normal plane, yet the calculations are carried out in the transverse plane. To work helical gear problems, all normal values must be transferred to the transverse plane prior to calculating. Quite often a series of gears is designed to achieve different ratios on the same center distance. Let us look at a helical gear example to illustrate the mathematics involved.

An electric motor operating at 3550 rpm drives a compressor. It is designed to operate the compressor at two different speeds, 33,897 and 31,842 rpm. With a 296-tooth gear driving, a 31-tooth pinion will achieve 33,897 rpm and a 33-tooth pinion will achieve 31,842 rpm. The idea is to use the same gear, housing, bearings, and so on, for both ratios, only changing the pinion to achieve either ratio. Also, it is desired to use 20 diametral pitch, 20° pressure angle cutters for all gears. The center distance is chosen by stress considerations as 8.4780 in. In order to encompass both the 31- and 33-tooth pinion designs, let us first calculate the gear geometry for a 32-tooth pinion design. The same gear will then mesh with the 31- and 33-tooth pinions.

The transverse diametral pitch of the 296 × 32 design is

$$DP_T \;=\; \frac{296 + 32}{2(8.4780)} \;=\; 19.34418495$$

The gear pitch diameter is

$$PD_G = \frac{296}{19.34418495} = 15.3017561$$

In order to use a standard 20 normal diametral pitch cutter the helix angle is

$$\psi = \cos^{-1} \frac{19.34418495}{20} = 14.71320405°$$

The gear lead is

$$L_G = \frac{\pi(15.3017561)}{\tan 14.71320405} = 183.0672333$$

For a normal pressure angle of $20°$ the transverse pressure angle is

$$\phi_T = \tan^{-1} \frac{\tan 20}{\cos 14.71320405} = 20.62180626°$$

For a standard gearset the pinion and gear outside diameter would be set by using a standard addendum of $1/DP$; therefore, the gear outside diameter would be

$$OD_G = 15.3017561 + \frac{2}{20} = 15.402$$

When a large gear is meshing with a small pinion it is conventional to increase the pinion addendum and decrease the gear addendum, resulting in what is commonly called a long and short addendum design. In the example the pinion addendum is 0.0629 and the gear addendum is 0.0405. These values are arrived at using two criteria:

1. The degrees of roll to the form diameter of the pinion in the 296 X 33 mesh must be high enough to avoid undercutting of the pinion. Undercutting occurs when the gear tooth tip describes an arc through space that would cut through the active profile of the pinion. In other words, the trochoid generated by the gear tooth tip would interfere with the pinion involute above the pinion form diameter.
2. The addendums are varied such that the temperature rise in the mesh due to sliding is minimized. This subject is discussed in Chapter 3.

It should be noted that the pinion addendum is lengthened by the same amount the gear addendum is shortened. Because of the long and short addendum design standard tooth thicknesses cannot be used, since this would result in an imbalance of bending strength between the pinion and gear, the pinion being weakened. Assuming a backlash in the mesh of 0.006, the standard transverse circular tooth thicknesses for pinion and gear would be

$$\text{CTTT}_P = \text{CTTT}_G = \frac{\text{CP}}{2} - \frac{\text{BL}}{2} = \frac{\pi}{2(19.34418495)} - \frac{0.006}{2}$$

$$= 0.078203$$

The optimized transverse circular tooth thicknesses for the 296 X 32 mesh ar are:

$$\text{CTTT}_P = 0.09072$$

$$\text{CTTT}_G = 0.06569$$

Now let us look at the 296 X 31 tooth mesh. The transverse diametral pitch is

$$\text{DP}_T = \frac{296 + 31}{2(8.4780)} = 19.28520878$$

and the gear pitch diameter is

$$\text{PD}_G = \frac{296}{19.28520878} = 15.34855046$$

The helix angle at this diameter is

$$\psi = \tan^{-1} \left(\frac{\pi(15.34855046)}{183.0672333} \right) = 14.75623793°$$

and the transverse pressure angle is

$$\phi_T = \cos^{-1} \left(\frac{\cos 20.62180626(15.3017561)}{15.34855046} \right) = 21.08111697°$$

Knowing the gear transverse circular tooth thickness at the 15.3017561 diameter (0.06569), we can calculate the gear transverse circular tooth thickness at the 15.34855046 diameter:

$$\text{CTTT}_G = 0.0480391$$

and to achieve a backlash of 0.006 the pinion transverse circular tooth thickness is

$$\text{CTTT}_P = \frac{\pi}{19.28520878} - 0.0480391 - 0.006 = 0.10886192$$

Since we cannot optimize the addendums for both the 296 X 31 and 296 X 33 tooth meshes we will use the pinion outside diameter calculated for the 296 X 32 mesh 1.780 for the 31- and 33-tooth pinions. Figures 2.32 and 2.33 are computer printouts giving all the tooth geometry for the 296 X 31 and 296 X 33 meshes.

296 X 31 MESH

AMERICAN LOHMANN GEAR

	DRIVEN PINION	DRIVER EXTERNAL GEAR
NUMBER OF TEETH	31.0000000	296.0000000
HELIX ANGLE (DEG)	14.7562380	14.7562380
PITCH DIAMETER	1.6074495	15.3485505
RELATIVE ROLLING SPEED (RPM)	33897.0000000	3550.0236486
MESH TORQUE (IN-LBS)	316.0825442	3018.0784863
BENDING GEOMETRY FACTOR	0.5000000	0.5000000
BENDING STRESS (PSI)	19677.4648288	19677.4648288
BENDING LIFE (HOURS)	999999.0000000	999999.0000000
BENDING SAFETY FACTOR	2.2360604	2.2360604
COMPRESSIVE STRESS (PSI)	84528.9062052	84528.9062052
COMPRESSIVE LIFE (HOURS)	999999.0000000	999999.0000000
COMPRESSIVE SAFETY FACTOR	1.6009171	1.6009171
SLIDING VELOCITY AT TIP (FPM)	3728.8957744	-931.9806035
A.G.M.A. MATERIAL GRADE	1.0000000	1.0000000
ALTERNATING BENDING FACTOR	1.0000000	1.0000000
NUMBER OF MESHES PER REV	1.0000000	1.0000000
BASE HELIX ANGLE (DEG)	13.8076961	13.8076961
OUTSIDE DIAMETER	1.777- 1.780	15.380-15.383
PITCH DIAMETER	1.6074495	15.3485505
FORM DIAMETER	1.5757543	15.2158791
BASE DIAMETER	1.4998664	14.3213053
ROOT DIAMETER	1.521- 1.531	15.124-15.134
ROLL ANGLE-MAX OUTSIDE DIA	36.6168247	22.4672304
ROLL ANGLE-ROUND EDGE DIA	36.0681676	22.3643649
ROLL ANGLE-HIGH SINGLE TOOTH	30.0682702	21.7814020
ROLL ANGLE-PITCH DIA	22.0869009	22.0869009
ROLL ANGLE-LOW SINGLE TOOTH	25.0039227	21.2510143
ROLL ANGLE-FORM DIAMETER	18.4553682	20.5651859
TOP LAND THICKNESS	0.0264463	0.0347370
MAX CASE DEPTH	0.0241853	0.0261853
TRANSVERSE CIR TOOTH THICKNESS	0.1088630	0.0480387
NORMAL CIR TOOTH THICKNESS	0.1048- 0.1058	0.0460- 0.0470
NORMAL DIAMETRAL PITCH	19.9429632	19.9429632
NORMAL PRESSURE ANGLE	20.4441987	20.4441987
LEAD	19.1725861	183.0672733
ROUND EDGE RADIUS MAX	0.0050000	0.0050000
ROOT FILLET RADIUS MIN	0.0201191	0.0206911
WHOLE DEPTH CONSTANT	2.4000000	2.4000000
CLEARANCE AT TIP OF TOOTH	0.0209483	0.0209483
BALL DIAMETER	0.1250000	0.1250000
MEASUREMENT OVER BALLS	1.9000- 1.901815	15.5315-15.5341
BALL CONTACT DIAMETER	1.7152- 1.716915	15.3622-15.3647
DIM OVER TOP LAND	0.0600097	0.0742515

TRANSVERSE DIAMETRAL PITCH	19.2852008
TRANSVERSE PRESSURE ANGLE(DEG)	21.0811130
CENTER DISTANCE	8.4780000
PITCH LINE VELOCITY (FPM)	14264.8838538
MESH RATIO	9.5483871
RELATIVE HORSEPOWER PER MESH	170.0000000
EFFECTIVE FACE WIDTH	1.6250000
STATIC TANGENTIAL LOAD (LBS)	393.2721197
DYNAMIC FACTOR	1.5349331
ALIGNMENT FACTOR	1.3733637
MODIFIED TANGENTIAL LOAD (LBS)	829.0260354
SURFACE FINISH	20.0000000
FLASH TEMPERATURE RISE (DEG F)	72.9815383
PROFILE CONTACT RATIO	1.5640263
FACE CONTACT RATIO	2.6274494
MIN CONTACT LENGTH	2.5137344
MAXIMUM CONTACT LENGTH	2.7510625
BACKLASH	0.0060000
CIRCULAR PITCH	0.1629017
LEAD ERROR (IN/IN)	0.0002000
BASE PITCH	0.1519990
DEPTH TO POINT OF MAX SHEAR	-0.0027174

Figure 2.32 A 296 X 31 tooth computer output sheet. (Courtesy of American Lohmann Corporation, Hillside, N.J.)

```
296 X 33 MESH
                                                          AMERICAN LOHMANN GEAR
                              DRIVEN        DRIVER
                              PINION     EXTERNAL GEAR

NUMBER OF TEETH               33.0000000   296.0000000    TRANSVERSE DIAMETRAL PITCH        19.4031611
HELIX ANGLE (DEG)            14.6704150    14.6704150     TRANSVERSE PRESSURE ANGLE(DEG)    20.1524970
PITCH DIAMETER                1.7007538    15.2552462     CENTER DISTANCE                    8.4790000

RELATIVE ROLLING SPEED (RPM) 31842.0000000  3549.9527027 PITCH LINE VELOCITY (FPM)      14177.9833990
MESH TORQUE (IN-LBS)         296.8956096   2663.0636495   MESH RATIO                         8.9696970
BENDING GEOMETRY FACTOR       0.5000000     0.5000000     RELATIVE HORSEPOWER PER MESH     150.0000000
BENDING STRESS (PSI)         19220.8571549  19220.8571549 EFFECTIVE FACE WIDTH               1.6250000
BENDING LIFE (HOURS)         999999.0000000 999999.0000000 STATIC TANGENTIAL LOAD (LBS)    349.1341424
BENDING SAFETY FACTOR         2.2891799     2.2891799     DYNAMIC FACTOR                     1.5316706
COMPRESSIVE STRESS (PSI)     75092.5963251  75092.5963251 ALIGNMENT FACTOR                  1.5051070
COMPRESSIVE LIFE (HOURS)     999999.0000000 999999.0000000 MODIFIED TANGENTIAL LOAD (LBS)  804.8660706
COMPRESSIVE SAFETY FACTOR     1.8110973     1.8110973     SURFACE FINISH                   20.0000000
SLIDING VELOCITY AT TIP (FPM) 1861.7184171 -3335.9694645  FLASH TEMPERATURE RISE (DEG F)    78.9349229

A.G.M.A. MATERIAL GRADE       1.0000000     1.0000000     PROFILE CONTACT RATIO             1.8454403
ALTERNATING BENDING FACTOR    1.0000000     1.0000000     FACE CONTACT RATIO                2.6274494
NUMBER OF MESHES PER REV      1.0000000     1.0000000     MIN CONTACT LENGTH                3.0514010
BASE HELIX ANGLE (DEG)       13.8076960    13.8076960     MAXIMUM CONTACT LENGTH            3.1498422

OUTSIDE DIAMETER            1.777- 1.780  15.380-15.383   BACKLASH                          0.0060000
PITCH DIAMETER                1.7007538    15.2552462     CIRCULAR PITCH                    0.1619114
FORM DIAMETER                 1.6125344    15.1871940     LEAD ERROR (IN/IN)                0.0002000
BASE DIAMETER                 1.5966320    14.3213052     BASE PITCH                        0.1519770
ROOT DIAMETER               1.523- 1.533  15.126-15.136   DEPTH TO POINT OF MAX SHEAR      0.0026780

ROLL ANGLE-MAX OUTSIDE DIA   28.2371869    22.4672316
ROLL ANGLE-ROUND EDGE DIA    27.5242396    22.3643661
ROLL ANGLE-HIGH SINGLE TOOTH 19.0158108    21.4391728
ROLL ANGLE-PITCH DIA         21.0268157    21.0268157
ROLL ANGLE-LOW SINGLE TOOTH  17.3280971    21.2510155
ROLL ANGLE-FORM DIAMETER      8.1067210    20.2229567

TOP LAND THICKNESS            0.0415479     0.0347377
MAX CASE DEPTH                0.0321538     0.0261538
TRANSVERSE CIR TOOTH THICKNESS 0.0731610    0.0827504
NORMAL CIR TOOTH THICKNESS  0.0703- 0.0713 0.0796- 0.0806
NORMAL DIAMETRAL PITCH       20.0570477    20.0570477
NORMAL PRESSURE ANGLE        19.5460270    19.5460270
LEAD                         20.4095271   183.0672734
ROUND EDGE RADIUS MAX         0.0050000     0.0050000
ROOT FILLET RADIUS MIN        0.0241296     0.0197955
WHOLE DEPTH CONSTANT          2.4000000     2.4000000
CLEARANCE AT TIP OF TOOTH     0.0201918     0.0201918

BALL DIAMETER                 0.1250000     0.1250000
MEASUREMENT OVER BALLS      1.9326- 1.934615.5315-15.5341
BALL CONTACT DIAMETER       1.7558- 1.757615.3622-15.3647
DIM OVER TOP LAND             0.0762863     0.0742512
```

Figure 2.33 A 296 × 33 tooth computer output sheet. (Courtesy of American Lohmann Corporation, Hillside, N.J.)

ENGINEERING DRAWING FORMAT

The engineering drawing must contain sufficient information to define the component completely so that the manufacturing department can fabricate it and the quality control department can inspect it. There are several elements that should appear on the field of a gear drawing:

1. Gear blank features are usually shown in an end view and cross section, as illustrated in Figure 2.34. It is important to specify the reference surfaces that will locate the gear in the application. For instance, the gear in Figure 2.34 will be pressed onto a shaft; therefore, the surface C locates the gear in the assembly and the gear tooth geometry must be accurate with respect to this surface. Sides A and B must be parallel to each other according to the drawing, and surface C perpendicular to A and B. The surface finish is designated by the $\nearrow$ symbol. The end view shows a gear tooth and calls out which face of the gear tooth is loaded. The X's indicate to how many decimal places a dimension is given.

2. A close-up view of the gear teeth as shown in Figure 2.35 defines the outside, pitch, form, and root diameters. It also calls out the roughness in the root area and in the area between the form diameter and the outside diameter: the active profile. The maximum undercut allowed in the root fillet

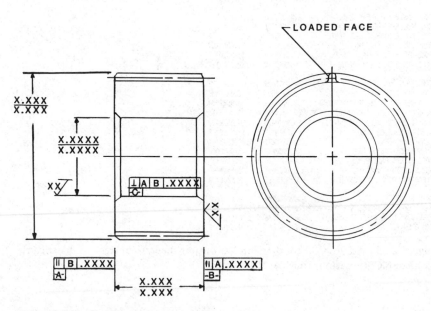

Figure 2.34   Gear blank dimensioning.

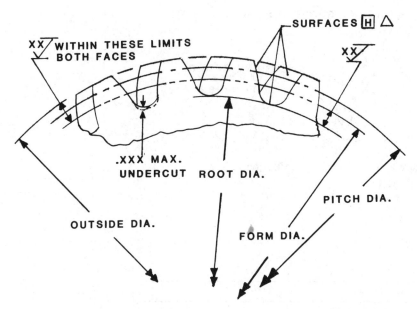

Figure 2.35 View of gear teeth.

area is defined. If the teeth are surface-hardened by processes such as carburizing or nitriding, the areas to be hardened are designated as surfaces [H]
followed by a triangle, which refers to a note that defines the case hardness
and depth. Note that in the illustration the top lands and tooth ends are not
hardened. Some designers prefer to harden these areas and therefore would
point to them in this view.

3. The tooth edges at the top land and the ends must be rounded, and Figure
2.36 illustrates how the radii of the tooth tips and edges are defined.
4. The gear material and its heat treatment must be specified and this is usually
done in a block of data on the lower right-hand side of the drawing. As an
illustration the callout for a carburized gear follows:

Material: AMS 6265 forging
Carburize surfaces [H]
Effective case depth of finished gear: 0.035 to 0.050
Case hardness: R_c 60 to 63
Core hardness: R_c 32 to 40
Per specification xxxx
Surface temper inspection per specification xxxx
Magnetic particle inspection per specification xxxx

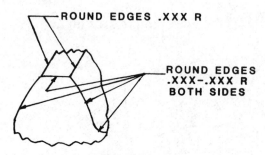

ROUND EDGES .XXX R

ROUND EDGES
.XXX–.XXX R
BOTH SIDES

Figure 2.36   Round edge definition.

GEAR DATA			
METHOD OF MANUFACTURE			
NO. OF TEETH			
HELIX ANGLE	XX.XXXX°		
HAND OF HELIX			
LEAD OF HELIX	XXX.XXXX		
NORMAL DIAMETRAL PITCH	XX.XXXX		
NORMAL PRESSURE ANGLE	XX.XXXX°		
NORMAL CIRC. TOOTH THICK.	.XXXX/.XXXX		
PITCH DIAMETER	XX.XXXX	⊙ C .XXXX	
ROOT DIAMETER	XX.XXX/XX.XXX		
FORM DIAMETER, MAX.	XX.XXXX		
WIRE OR BALL DIAMETER	.XXXX		
MEAS. OVER WIRE OR BALLS	XX.XXXX/XX.XXXX		
AGMA QUALITY NUMBER			
RUN OUT TOL.	.XXX		
PITCH TOL.	.XXXX		
PROFILE TOL.	SEE DIAGRAM		
LEAD TOL.	SEE DIAGRAM		
MATING GEAR PART NUMBER			
BACKLASH WITH MATING GEAR			

Figure 2.37   Gear data block.

5. Figure 2.37 is one form of a gear data block which is applicable to both spur and helical gears. In this format all data are given at the operating pitch diameter. Note that the pitch diameter must be concentric to surface C shown in Figure 2.34.

Tooth Tolerances

Note that the gear data form (Figure 2.37) has a line which calls out the AGMA quality number. AGMA Standard 390.03 [1] specifies quality numbers identifying specific tooth element tolerances. The higher the quality number, the more precise the gearing will be and the closer the tolerances. Quality numbers range from 3 to 15 and the standard contains a tabulation of many industrial and end use applications and suggested quality number ranges for each. The following table lists some sample applications:

Industry	Quality number
Aerospace engines	10-13
Agriculture	3-7
Automotive	10-11
Mining	5-8
Steel	5-6

In the industries cited above and in other applications, when high-speed drives are required or there are special considerations such as noise abatement, higher quality numbers may be called for. It should be noted that quality classes 13, 14, and 15 are extremely difficult to achieve and prior to requiring these classes there should be agreement between the manufacturer and user as to the method of inspection.

The majority of critical industrial applications in fields such as the process industries and turbomachinery will require gear units with elements that fall into the quality number range 10 to 13. Table 2.1 presents the tolerances for these classes. Following is a definition of each tooth tolerance element shown in Table 2.1:

1. *Runout tolerance.* The variation of the pitch diameter in a direction perpendicular to the axis of rotation with respect to a reference surface of revolution such as a bearing journal or a bore. The pitch diameter, being theoretical, must be indirectly measured and this can be done several ways. Two widely used methods are:
   a. Runout can be measured by indicating the position of a ball probe in successive teeth (see Figure 2.38).

Table 2.1 Pitch Gear Tolerances (in ten-thousandths of an inch)

AGMA QUALITY NUMBER	NORMAL DIAMETRAL PITCH	RUNOUT TOLERANCE — PITCH DIAMETER (INCHES)										PITCH — PITCH DIAMETER				
		3/4	1½	3	6	12	25	50	100	200	400	3/4	1½	3	6	12
8	1/2					146.5	174.5	205.8	242.7	286.3	337.6					19.0
	1				88.8	104.8	124.8	147.2	173.6	204.7	241.4				14.4	16.3
	2			53.9	63.5	74.9	89.2	105.2	124.1	146.3	172.6			10.9	12.3	14.0
	4		32.7	38.5	45.4	53.6	63.8	75.2	88.7	104.6	123.4		8.3	9.3	10.6	11.9
	8	19.8	23.3	27.5	32.5	38.3	45.6	53.8	63.4	74.8	88.2	6.3	7.1	8.0	9.0	10.2
	12	16.3	19.2	22.6	26.7	31.5	37.5	44.2	52.1	61.5	72.5	5.7	6.5	7.3	8.3	9.3
	20	12.7	15.0	17.7	20.8	24.6	29.3	34.5	40.7	48.0	56.6	5.1	5.8	6.5	7.4	8.3
9	1/2					104.7	124.7	147.0	173.4	204.5	241.2					13.4
	1				63.5	74.8	89.1	105.1	124.0	146.2	172.4				10.2	11.5
	2			38.5	45.4	53.5	63.7	75.2	88.6	104.5	123.3			7.7	8.7	9.8
	4		23.3	27.5	32.4	38.3	45.6	53.7	63.4	74.7	88.1		5.8	6.6	7.4	8.4
	8	14.1	16.7	19.7	23.2	27.4	32.6	38.4	45.3	53.4	63.0	4.4	5.0	5.6	6.4	7.2
	12	11.6	13.7	16.2	19.1	22.5	26.8	31.6	37.2	43.9	51.8	4.0	4.6	5.1	5.8	6.6
	20	9.1	10.7	12.6	14.9	17.6	20.9	24.7	29.1	34.3	40.4	3.6	4.1	4.6	5.2	5.9
10	1/2					74.8	89.0	105.0	123.8	146.1	172.3					9.4
	1				45.3	53.5	63.7	75.1	88.5	104.4	123.2				7.2	8.1
	2			27.5	32.4	38.2	45.5	53.7	63.3	74.7	88.1			5.4	6.1	6.9
	4		16.7	19.6	23.2	27.3	32.5	38.4	45.3	53.4	63.0		4.1	4.6	5.2	5.9
	8	10.1	11.9	14.0	16.6	19.5	23.3	27.4	32.4	38.2	45.0	3.1	3.5	4.0	4.5	5.1
	12	8.3	9.8	11.5	13.6	16.1	19.1	22.6	26.6	31.4	37.0	2.8	3.2	3.6	4.1	4.6
	20	6.5	7.6	9.0	10.6	12.5	14.9	17.6	20.8	24.5	28.9	2.5	2.9	3.2	3.7	4.1
11	1/2					53.4	63.6	75.0	88.5	104.3	123.0					6.6
	1				32.4	38.2	45.5	53.6	63.2	74.6	88.0				5.0	5.7
	2			19.6	23.1	27.3	32.5	38.3	45.2	53.3	62.9			3.8	4.3	4.9
	4		11.9	14.0	16.6	19.5	23.2	27.4	32.3	38.1	45.0		2.9	3.3	3.7	4.2
	8	7.2	8.5	10.0	11.8	14.0	16.6	19.6	23.1	27.3	32.2	2.2	2.5	2.8	3.2	3.6
	12	5.9	7.0	8.2	9.7	11.5	13.7	16.1	19.0	22.4	26.4	2.0	2.3	2.6	2.9	3.3
	20	4.6	5.5	6.4	7.6	9.0	10.7	12.6	14.8	17.5	20.6	1.8	2.0	2.3	2.6	2.9
12	1/2					38.1	45.4	5.36	63.2	74.5	87.9					4.7
	1				23.1	27.3	32.5	38.3	45.2	53.3	62.8				3.5	4.0
	2			14.0	16.5	19.5	23.2	27.4	32.3	38.1	44.9			2.7	3.0	3.4
	4		8.5	10.0	11.8	13.9	16.6	19.6	23.1	27.2	32.1		2.0	2.3	2.6	2.9
	8	5.2	6.1	7.2	8.5	10.0	11.9	14.0	16.5	19.5	23.0	1.5	1.7	2.0	2.2	2.5
	12	4.2	5.0	5.9	6.9	8.2	9.8	11.5	13.6	16.0	18.9	1.4	1.6	1.8	2.0	2.3
	20	3.3	3.9	4.6	5.4	6.4	7.6	9.0	10.6	12.5	14.7	1.3	1.4	1.6	1.8	2.0

Source: AGMA Standard 390.03.

TOLERANCE (INCHES)					PROFILE TOLERANCE — PITCH DIAMETER (INCHES)										LEAD TOLERANCE — FACE WIDTH (INCHES)				
25	50	100	200	400	3/4	1½	3	6	12	25	50	100	200	400	1 and Less	2	3	4	5
21.7	24.5	27.7	31.3	35.4					42.6	47.7	53.1	59.1	65.7	73.1					
18.6	21.0	23.7	26.8	30.3				28.3	31.5	35.3	39.3	43.7	48.6	54.1					
15.9	18.0	20.3	23.0	26.0			18.8	21.0	23.3	26.1	29.0	32.3	36.0	40.0					
13.6	15.4	17.4	19.7	22.2		12.5	13.9	15.5	17.2	19.3	21.5	23.9	26.6	29.6	5	8	11	13	16
11.7	13.2	14.9	16.8	19.0	8.3	9.3	10.3	11.5	12.8	14.3	15.9	17.7	19.7	21.9					
10.6	12.0	13.6	15.4	17.4	7.0	7.8	8.6	9.6	10.7	12.0	13.3	14.8	16.5	18.4					
9.5	10.7	12.1	13.7	15.5	5.6	6.2	6.9	7.7	8.6	9.6	10.7	11.9	13.2	14.7					
15.3	17.3	19.5	22.1	24.9					30.4	34.1	37.9	42.2	46.9	52.2					
13.1	14.8	16.7	18.9	21.4				20.2	22.5	25.2	28.1	31.2	34.7	38.6					
11.2	12.7	14.3	16.2	18.3			13.5	15.0	16.7	18.6	20.7	23.1	25.7	28.6					
9.6	10.8	12.2	13.8	15.7		8.9	10.0	11.1	12.3	13.8	15.3	17.1	19.0	21.1	4	7	9	11	13
8.2	9.3	10.5	11.9	13.4	5.9	6.6	7.4	8.2	9.1	10.2	11.4	12.6	14.1	15.6					
7.5	8.5	9.6	10.8	12.2	5.0	5.5	6.2	6.9	7.6	8.6	9.5	10.6	11.8	13.1					
6.7	7.6	8.5	9.7	10.9	4.0	4.4	4.9	5.5	6.1	6.8	7.6	8.5	9.4	10.5					
10.8	12.2	13.7	15.5	17.6					21.7	24.3	27.1	30.1	33.5	37.3					
9.2	10.4	11.8	13.3	15.0				14.5	16.1	18.0	20.0	22.3	24.8	27.6					
7.9	8.9	10.1	11.4	12.9			9.6	10.7	11.9	13.3	14.8	16.5	18.3	20.4					
6.7	7.6	8.6	9.8	11.0		6.4	7.1	7.9	8.8	9.9	11.0	12.2	13.6	15.1	3	5	7	9	10
5.8	6.5	7.4	8.3	9.4	4.2	4.7	5.3	5.9	6.5	7.3	8.1	9.0	10.0	11.2					
5.3	6.0	6.7	7.6	8.6	3.6	4.0	4.4	4.9	5.5	6.1	6.8	7.6	8.4	9.4					
4.7	5.3	6.0	6.8	7.7	2.9	3.2	3.5	3.9	4.4	4.9	5.4	6.1	6.7	7.5					
7.6	8.6	9.7	10.9	12.4					15.5	17.4	19.3	21.5	24.0	26.7					
6.5	7.3	8.3	9.4	10.6				10.3	11.5	12.9	14.3	15.9	17.7	19.7					
5.6	6.3	7.1	8.0	9.1			6.9	7.6	8.5	9.5	10.6	11.8	13.1	14.6					
4.8	5.4	6.1	6.9	7.8		4.6	5.1	5.6	6.3	7.0	7.8	8.7	9.7	10.8	3	4	6	7	8
4.1	4.6	5.2	5.9	6.6	3.0	3.4	3.8	4.2	4.6	5.2	5.8	6.4	7.2	8.0					
3.7	4.2	4.7	5.4	6.1	2.5	2.8	3.1	3.5	3.9	4.4	4.9	5.4	6.0	6.7					
3.3	3.7	4.2	4.8	5.4	2.0	2.3	2.5	2.8	3.1	3.5	3.9	4.3	4.8	5.4					
5.3	6.0	6.8	7.7	8.7					11.1	12.4	13.8	15.4	17.1	19.0					
4.6	5.2	5.8	6.6	7.5				7.4	8.2	9.2	10.2	11.4	12.7	14.1					
3.9	4.4	5.0	5.6	6.4			4.9	5.5	6.1	6.8	7.6	8.4	9.4	10.4					
3.3	3.8	4.3	4.8	5.5		3.3	3.6	4.0	4.5	5.0	5.6	6.2	6.9	7.7	2	3	5	6	7
2.9	3.2	3.7	4.1	4.7	2.2	2.4	2.7	3.0	3.3	3.7	4.1	4.6	5.1	5.7					
2.6	3.0	3.3	3.8	4.3	1.8	2.0	2.2	2.5	2.8	3.1	3.5	3.9	4.3	4.8					
2.3	2.6	3.0	3.4	3.8	1.5	1.6	1.8	2.0	2.2	2.5	2.8	3.1	3.4	3.8					

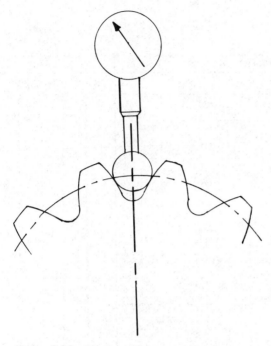

Figure 2.38 Single probe runout check.

  b. a rolling check can be conducted meshing the gear to be inspected with a
    master gear of known accuracy on a fixture with a movable center dis-
    tance. The variation of center distance is a measure of runout.
2. *Pitch tolerance.* The pitch is the theoretical distance between corresponding
  points on adjacent teeth. The variation from tooth to tooth can be measured
  using an instrument which employs a fixed finger and stop for consistent
  positioning on successive pairs of teeth, and a movable finger which displays
  pitch variations on a dial indicator or chart recorder (Figure 2.39).
3. *Profile tolerance.* The deviation from a true involute checked on an involute profile measuring instrument. In most cases, a modified involute is
  used; the drawing specification for involute modifications is discussed later
  in the chapter.
4. *Lead tolerance.* For a spur gear the lead inspection might be considered a
  check of the parallelism of the tooth with respect to the axis of rotation.
  The lead of a helical gear is the axial advance of a helix for one complete
  turn. Lead is checked by an instrument that advances a probe along the
  tooth surface, parallel to the axis, while the gear rotates in a specified, timed
  relation based on the lead.

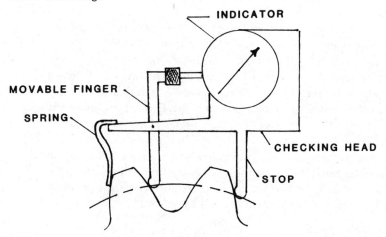

Figure 2.39 Tooth-to-tooth spacing check.

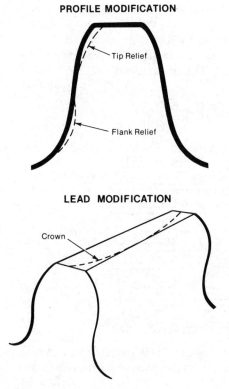

Figure 2.40 Tooth modifications.

In the data block (Figure 2.37) there are lines available for both the AGMA quality number and the specific tolerances mentioned above. It is possible that for a specific application the designer will not choose tolerances from a single quality number class. For instance, the designer may want to have a closer tolerance on tooth-to-tooth spacing that on profile. In such a case the individual tolerances can be specified on the gear data block. Even when a single quality number is used, the tolerances can be placed on the block for reference.

When a modified involute or lead is required, a note in the gear data block will refer to a diagram on the drawing which defines the modification. Modified involute profiles and leads are used to attempt to compensate for deflections during operation and tooth errors. Figure 2.40 illustrates profile (involute) modifications and lead modification, sometimes called crowning.

Profile Modification

Tooth profiles are modified to avoid interference which can occur as the teeth enter into or leave the mesh. The interference is a result of deflection of the gear teeth, shafts, or gear casing due to the transmitted load or tooth discrepancies such as spacing or profile error. For instance, if a pinion tooth is misplaced from its theoretical position due to spacing error or because the previous tooth has deflected under load and enters into mesh too soon, the interference with the mating tooth will create a dynamic load which will increase tooth stresses and system noise and vibration levels. Such interference can be eliminated by relieving the pinion and gear tooth tips or flanks or both, as shown in Figure 2.40. Figure 2.41 illustrates how the profile modification is specified on the engineering

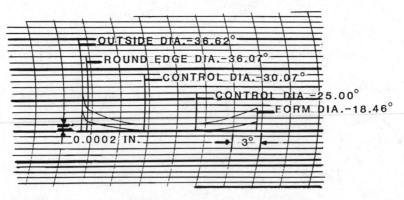

Figure 2.41 Modified involute diagram. The gear tooth profiles within the tolerance bands shall not depart from a smooth and gradual convex curvature.

drawing. The diagram provides tolerance bands for the chart that will be produced when the gear is inspected on an involute measuring instrument. Figure 2.41 presents a diagram for the 31-tooth pinion described in Figure 2.32. The amount of relief specified at the round edge diameter is 0.0003 to 0.0007. It is impractical to specify values at the outside diameter since the round edge radius cannot be closely controlled. The relief at the form diameter in this case is also 0.0003 to 0.0007. The tip and flank reliefs commence at the highest and lowest points of single tooth contact.

It should be recognized that the benefits from profile modification can be realized only if the teeth are accurately manufactured to tolerances less than the modifications specified. It would be pointless to have a 0.0005 tip modification if the tooth spacing were allowed to vary by 0.001. An estimate of the amount of modification required for a given application may be found in Ref. 2.

The modification at the first point of contact is given as:

$$\text{Modification} = \text{driving load (lb)} \times \frac{3.5 \times 10^{-7}}{\text{face width (in.)}}$$

To achieve this modification, material must be removed from the tip of the driven gear or the flank of the driving gear or both. If material is removed from both, the total modification is split between the two meshing teeth.

The modification at the last point of contact is given as:

$$\text{Modification} = \text{driving load (lb)} \times \frac{2.0 \times 10^{-7}}{\text{face width (in.)}}$$

To achieve this modification, material must be removed from the tip of the driving gear or the flank of the driven gear or both.

The foregoing estimates for profile modification are offered as starting points for a design. The final tooth modifications are arrived at through development by observing operating results.

Lightly loaded gears may not require profile modification. If a simple involute tolerance is called out on the gear data block, Figure 2.42 shows how this tolerance is to be interpreted. Assume a profile tolerance of 0.0008 in. The measured profile must fall within the checked area of the diagram in Figure 2.42. A true involute would be a straight line on the diagram. If the involute is plus, the line on the diagram will slant up toward the left. A minus involute is depicted by a dashed line on the diagram. In general, a plus involute tends toward an interference condition; therefore, the minus involute is more desirable.

Lead (Axial) Modifications

In theory, when gear teeth mesh the faces will be parallel to each other and the load will be distributed across the full face width. In practice, however, a

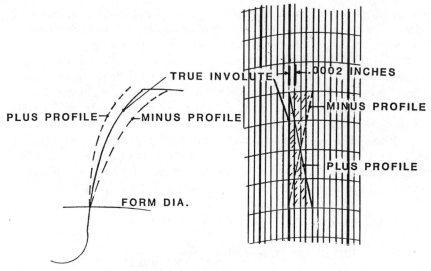

Figure 2.42 Interpretation of involute tolerances.

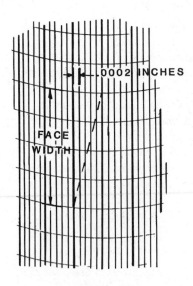

Figure 2.43 Lead diagram.

tolerance must be given. For spur gears it will be a tolerance on the parallelism of the tooth with the axis of rotation. For helical gears it is called a lead tolerance. Figure 2.43 shows how this tolerance is interpreted. The solid line is the theoretical trace and the dashed line is the measurement as recorded by a lead checking instrument. The diagram shows a 0.001-in. variation. If this is a right-hand helical gear, because the variation is off to the right, the measured helix angle is greater than the theoretical value.

Tooth faces tend not to be exactly parallel in operation not only because of tooth errors but also due to deflections of shafts, bearings, and casings. The load, therefore, may be concentrated on an end rather than distributed evenly across the face. To alleviate end loading, a lead modification, sometimes called a crown, is used. As shown in Figure 2.40, the crown relieves the tooth ends and avoids a heavy concentration of load in these areas. Figure 2.44 illustrates how a crowned tooth is specified on the engineering drawing. In this example the relief at either end of the face width is 0.0004 to 0.0008, blending smoothly into the flat at the center of the tooth. The amount of crowning generally is on the order of 0.001 in., but like the profile modification must be finally developed by observation of test results. In some cases deflections are such that only one end of the tooth need be crowned. On occasion two mating gears are designed with slightly different helix angles which become parallel as the system deflects under load.

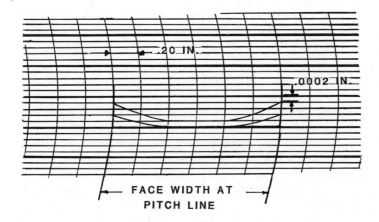

**FACE WIDTH AT
PITCH LINE**

Figure 2.44 Crowned lead diagram. The lead contour within the tolerance bands shall not depart from a smooth and gradual convex curvature.

SPLINE DESIGN

Splines are used in mechanical systems to transmit torque and motion from one shaft to another. A spline connection consists of a set of external gear teeth arranged in a circle which fit into a corresponding set of internal gear teeth. Splines provide a strong, compact method of connection which can accommodate some misalignment.

In general use today are involute splines of 30° pressure angle with stub teeth. Stub teeth have short addendums of 1/2(diametral pitch) rather than the conventional 1/diametral pitch. Because of this the spline pitch is conventionally given as a fraction (e.g., 12/24, the numerator being the diametral pitch, which is the number of spline teeth per inch of pitch diameter and which controls the circular pitch and basic space width or tooth thickness. The denominator is known as the stub pitch and is always twice the numerator. The tooth addendum is 1/stub pitch.

Spline teeth are of the involute form because of tooling advantages. A single hob or shaper cutter can generate all numbers of teeth of a given diametral pitch. Stub teeth and 30° pressure angles are used for ease of machining. The relatively high pressure angle increases tool life because the tool has more clearance behind the cutting edge. Also, higher cutting speeds are possible. The stub tooth is advantageous for broaching internal splines and for rolling of teeth.

Generally, splines are designed to ANSI Standard B92.1a [3]. In addition to 30° pressure angles the standard presents dimensioning systems for 37.5° and 45° pressure angle splines, which are sometimes known as serrations. Figure 2.45 taken from the standard shows spline tooth nomenclature and how spline data are presented on the engineering drawing.

Two root fillet configurations are possible, as shown in Figure 2.45. One is the flat root spline, in which fillets join the arcs of major or minor circles to the tooth sides. The other is the full fillet root spline, in which a single fillet in the general form of an arc joins the sides of adjacent teeth. The full fillet root form is stronger and should be used if appreciable torque is transmitted through the spline.

There are two types of fits possible with mating splines. One is a side fit where the mating members contact on the sides of the teeth only and there is clearance at the major diameters. When using a side fit spline if more accurate centralization of the shafts is desired, this can be accomplished by the use of shaft shoulders, as shown in Figure 2.46. It is also possible to have a major diameter fit where the mating members contact at their major diameters and the tooth sides act only as drivers. In this case the standard provides for increased clearance at the sides to ensure that all centering will be at the major diameters.

To be sure that two mating splines will fit together with minimum clearance, the concept of effective and actual tooth space and tooth thickness

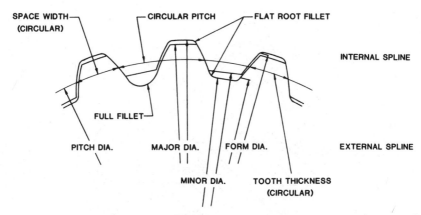

Figure 2.45 Spline nomenclature and drawing data.

DRAWING DATA				
INTERNAL INVOLUTE SPLINE DATA		**EXTERNAL INVOLUTE SPLINE DATA**		
FILLET ROOT SIDE FIT		FILLET ROOT SIDE FIT		
NUMBER OF TEETH	XX	NUMBER OF TEETH	XX	
SPLINE PITCH	XX/XX	SPLINE PITCH	XX/XX	
PRESSURE ANGLE	30°	PRESSURE ANGLE	30°	
BASE DIAMETER	X.XXXXXX REF.	BASE DIAMETER	X.XXXXXX REF.	
PITCH DIAMETER	X.XXXXXX REF.	PITCH DIAMETER	X.XXXXXX REF.	
MAJOR DIAMETER	X.XXX MAX.	MAJOR DIAMETER	X.XXX/X.XXX	
FORM DIAMETER	X.XXX	FORM DIAMETER	X.XXX	
MINOR DIAMETER	X.XXX/X.XXX	MINOR DIAMETER	X.XXX MIN.	
CIRCULAR SPACE WIDTH		CIRCULAR TOOTH THICKNESS		
MAX ACTUAL	X.XXXX	MAX EFFECTIVE	X.XXXX	
MIN EFFECTIVE	X.XXXX	MIN ACTUAL	X.XXXX	
MAX MEAS.BETW. PINS	X.XXXX REF.	MIN MEAS. OVER PINS	X.XXXX REF.	
PIN DIAMETER	X.XXXX	PIN DIAMETER	X.XXXX	

dimensions is used in spline tooth dimension systems. To understand this concept, imagine an internal spline with each tooth space width exactly half a circular pitch and the mating external spline with each tooth thickness exactly half a circular pitch. It would seem that these splines would fit perfectly; however, because of such tooth errors as spacing, profile, out of round, and lead, the pair probably cannot be assembled. Because of these errors the spline teeth will not be in their theoretical locations on the pitch circle and at some point or points there will be interference between the internal and external teeth. To overcome this problem, all space widths of the internal spline must be widened by the amount of interference caused by tooth errors and all tooth thicknesses

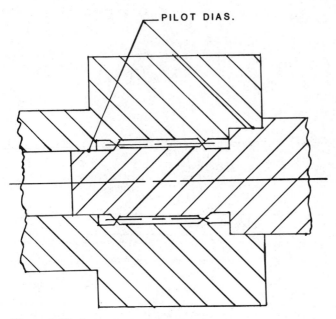

Figure 2.46 Piloted side fit spline.

of the external spline must be thinned. This concept leads to four dimensions for space width and tooth thickness:

Minimum effective space width = ½ circular pitch ($\phi = 30°$)
Maximum effective space width
Minimum actual space width
Maximum actual space width
Maximum effective tooth thickness = ½ circular pitch ($\phi = 30°$)
Minimum effective tooth thickness
Maximum actual tooth thickness
Minimum actual tooth thickness

The spline teeth are machined to the actual space width or tooth thickness dimensions which can be checked by the use of gages or measurements over pins. The effective dimensions are checked by gages. There are four machining tolerance classes set up for the effective and actual space widths and tooth thicknesses which result in varying degrees of clearance.

It must be remembered that the ability to assemble the spline is not the only criterion in critical applications. When significant loads are transmitted, or at high speed, the tooth geometry of the splined connection may have to be closely controlled and elements such as profile, lead, and surface texture specified.

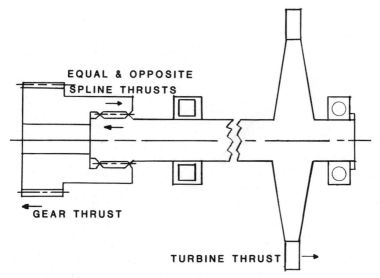

Figure 2.47 Helical spline shaft system.

In some cases where tooth bearing surface is important it may be desirable to use full-depth teeth. Full-depth splines would not use a 30° pressure angle since the teeth would be too pointed but would have a conventional 20° pressure angle or less.

There are applications where it is desirable to transmit thrust through a splined connection. For instance, a turbine wheel may be connected to a helical gear and the thrust of the wheel offset by the gear thrust. In such a case a helical tooth spline is effective. Figure 2.47 illustrates such a system. The spline helix angle is chosen such that the spline thrust exceeds the gear thrust and the shaft system locks up with the turbine shaft, bottoming out in the gear shaft shoulder. The net thrust in the system is then the turbine thrust minus the gear thrust, which is reacted by the ball bearing. The bearing loading, therefore, is greatly reduced from the case where the thrusts are not offset.

REFERENCES

1. AGMA Gear Handbook 390.03, Vol. 1, Gear Classification, Materials and Measuring Methods for Unassembled Gears, American Gear Manufacturers Association, Arlington, Va., January 1973.
2. Dudley, D. W., *Gear Handbook*, McGraw-Hill, New York, 1962, pp. 5–23.
3. ANSI Standard B92.1a, Involute Splines and Inspection, Society of Automotive Engineers, Warrendale, Pa., 1976.

3
GEARBOX RATING

The rating of a gearbox is determined by the loads the gearbox components are capable of transmitting. In some cases where a system is operating continuously at a uniform load such as an electric motor driving a fan, the loading is simple to predict and component analysis can be based on a continuous horsepower transmitted at steady speed. Some applications, however, experience variable loading such as high starting torque or shock loads and these conditions must be considered in the gearbox design.

Organizations such as the American Gear Manufacturers Association (AGMA) and the American Petroleum Institute (API) issue Standards that define gear rating procedures. AGMA Standard 420.04 [1] covers enclosed drives with pitch line velocities not exceeding 5000 fpm or pinion speeds not exceeding 3600 rpm. Higher-speed enclosed drives are covered by AGMA Standard 421.06 [2]. The general AGMA Standard for gear rating is 218.01 [4].

The rating methods used in these standards are discussed in this chapter. Before going into detail, an overview of the procedure one would use in rating a gearbox follows:

1. *Gear tooth rating.* The first step in determining a gearbox rating is to evaluate the tooth meshes. The classical gear tooth limitations that are calculated are the fatigue phenomena of breakage and pitting. Tooth breakage is analyzed by calculating the bending stress in the root fillet area and comparing it against a material strength rating. Pitting is analyzed by calculating the compressive stress at the tooth contact and comparing it against a material durability rating. A third gear tooth limitation encountered in high-speed gearing is instability of the lubricant film, allowing metal-to-metal contact leading to

scoring. The failure modes of tooth breakage, pitting, and scoring are described in Chapter 12. Their analysis is covered later in this chapter.

2. *Bearing rating*. Bearing ratings may be the limiting factor in determining the load a gearbox can transmit. A decision must be made as to the minimum acceptable L_{10} life desired for antifriction bearings or the maximum loading acceptable for journal bearings. The analysis and rating of bearings is presented in Chapter 4.

3. *Thermal rating*. Gear stresses or bearing lives usually determine the mechanical rating of a gearbox. In addition to the mechanical rating, gearboxes which do not use external cooling have a thermal capacity. This is defined in AGMA Standard 420.04 as the horsepower a unit will transmit continuously for 3 hr or more without exceeding a sump temperature of 200°F or a sump temperature rise of 100°F over ambient. If these thermal limits are exceeded, external cooling must be provided. Gearbox thermal ratings and lubrication systems are discussed in Chapter 5.

4. *Shaft rating*. Consideration must be given to gearbox components other than gears and bearings. Shafting, keyways, splines, and so on, must be analyzed to assure satisfactory performance under load. These machine elements are discussed later in the chapter.

The four points above are the obvious design details that must be addressed; however, there are many other details, such as housing and shaft deflections, critical speeds, and thermal expansion, of which the experienced gearbox designer is aware. Generally, in the procurement of a gearbox, gear ratings, bearing lives, and lubrication details are documented and the finer points of gearbox design are left to the manufacturer.

TOOTH LOADS

To calculate gear and shaft stresses and bearing lives, the gear tooth loads must be developed. Figure 3.1 illustrates the load diagram on a spur gear tooth. The gear torque is

$$T_q = \frac{63,025 \text{ hp}}{\text{rpm}}$$

where

T_q = torque, in.-lb
hp = horsepower
rpm = gear rotational speed

The total transmitted tooth load R acts normal to the involute profile. The component of R transmitting the torque is the tangential load:

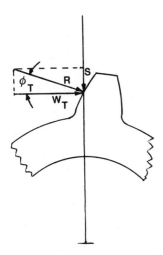

Figure 3.1 Spur gear tooth loads.

$$W_T = \frac{2T_q}{PD}$$

where

W_T = tangential load, lb
PD = gear pitch diameter, in.

The total transmitted load is

$$R = \frac{W_T}{\cos \phi_T}$$

where

R = total transmitted (resultant) force, lb
ϕ_T = transverse pressure angle, deg

As shown in Figure 3.1, the force R is resolved into the tangential load and a separating load:

$$S = W_T \tan \phi$$

where S is the separating load, in pounds. In the case of helical gears, the resultant force R is in the normal plane. To resolve R into a tangential and separating force, the geometry shown in Figure 3.2 is used. There is also a thrust force generated since the resultant force is at an angle ψ to the tangential force:

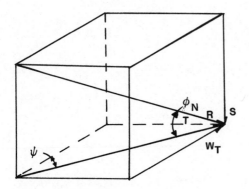

Figure 3.2   Helical gear tooth loads.

$$W_T = R \cos \phi_N \cos \psi$$

$$S = R \sin \phi_N = W_T \tan \phi_T$$

$$T = R \cos \phi_N \sin \psi = W_T \tan \psi$$

where

ψ = helix angle, deg
ϕ_N = normal pressure angle, deg
T = thrust load, lb

STRENGTH RATING

The strength rating of a gear tooth concerns itself with the bending stress (Figure 3.3) in the tooth fillet, where fatigue cracks initiate and propogate resulting in

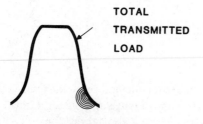

TOTAL TRANSMITTED LOAD

BENDING STRESS

Figure 3.3   Bending stress criterion for strength rating.

fracture of teeth or portions of teeth. The fundamental equation for bending stress in a gear tooth is [4]

$$S_t = \frac{W_T P_d}{FJ}$$

where

S_t = tensile or bending stress, psi
W_T = transmitted tangential load, lb
P_d = transverse diametral pitch, in.$^{-1}$
J = geometry factor

The geometry factor J is an index of the following:

Tooth geometry in the root fillet area
Stress concentration in the root fillet
Load sharing between teeth
The position at which the most damaging load is applied

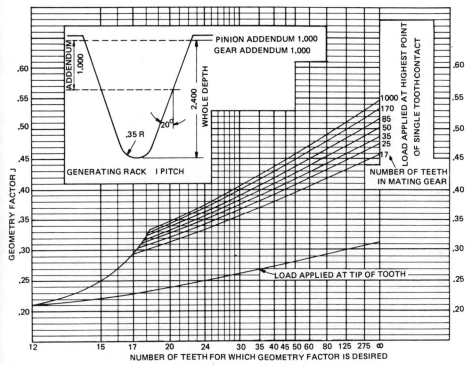

Figure 3.4a Geometry factors, 20° spur gears, standard addendum (From Ref. 5.)

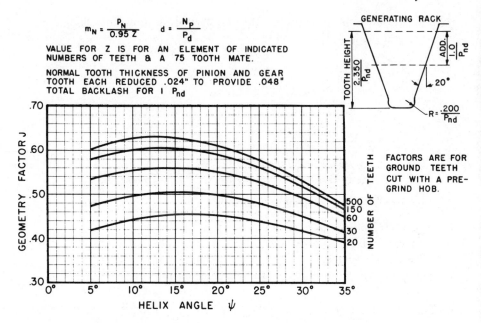

Figure 3.4b   Geometry factors, 20° helical gears, standard addendum. (From Ref. 5.)

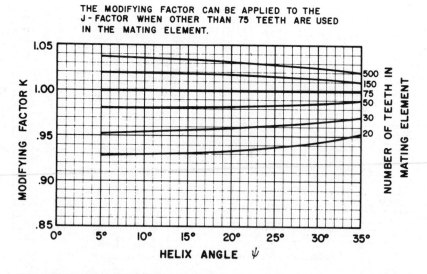

Figure 3.4c   Modifying factor for helical gear geometry factors. (From Ref. 5.)

Geometry factors may be arrived at by graphical layout of the tooth form or computer analysis of the graphical procedure. Reference 5 presents a method for calculating the geometry factor. Figures 3.4a, 3.4b, and 3.4c present geometry factors for $20°$ pressure angle spur and helical gears. The geometry factor is strongly dependent on the cutting tool geometry and these figures are for hobbed gears. In general, spur gear geometry factors vary from approximately 0.35 to 0.45 and helical gear geometry factors have values from approximately 0.4 to 0.6.

BENDING STRESS RATING

In the AGMA rating system the basic bending stress is modified by several factors that deal with the characteristics of a specific application:

$$S_t = \frac{W_T K_a}{K_v} \frac{P_d}{F} \frac{K_s K_m}{J} \tag{3.1}$$

where

K_a = application factor. This factor takes into account the roughness or smoothness of the driving and driven equipment. When no overloads are anticipated K_a may be taken as 1.0. For very rough operation K_a may be 2.25 or greater.

K_v = dynamic factor. The dynamic factor represents the ratio between the maximum dynamic load on the gear teeth and the static calculated load. Gear teeth generate dynamic loads due to component geometry errors which result in gear accelerations and decelerations. Although the dynamic factor is used as a multiplier in the stress equation, the dynamic load is actually an incremental force which adds to the tangential force.

The dynamic factor increases with increasing pitch line velocity and decreases with increasing tooth accuracy and increased tooth loading. As the tooth loading is increased, the tooth deflections tend to overshadow tooth geometry errors and the dynamic load is a smaller percentage of the total load. Figure 3.5 illustrates this trend. The data shown were developed from a test conducted on a helical planetary gearset transmitting 1100 hp at 21,000 rpm-in. The sun gear was strain gaged in the root to measure tooth loading. Gear quality was AGMA Quality Class 12 [6]. For gears of lower quality classes operating at lower speeds, the following estimates can be used for dynamic factors [4]:

$$K_v = \left(\frac{92}{92 + PLV} \right)^{0.25}$$

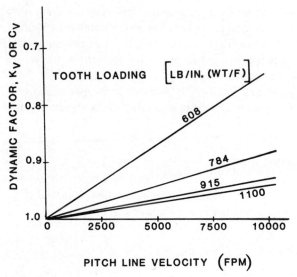

Figure 3.5 Dynamic factors for accurate gearing.

where PLV is the pitch line velocity in fpm, for AGMA Quality Class 11 gearing operating at pitch line velocities less than 8000 fpm with rigid accurate mountings;

$$K_v = \left(\frac{84}{84 + PLV}\right)^{0.4}$$

for AGMA Quality Class 10 gearing operating at pitch line velocities of less than 6000 fpm and

$$K_v = \left(\frac{70.7}{70.7 + PLV}\right)^{0.63}$$

for AGMA Quality Class 8 gearing operating at pitch line velocities of less than 5000 fpm.

There are analytical methods for calculating dynamic loads [7]. The tooth stiffness and mass are determined, and assuming the magnitude of tooth errors, an estimate of the dynamic load can be arrived at.

K_s = size factor. The size factor reflects nonuniformity of material properties which become more prevalent as the size of a gear increases;
however, standard size factors have not yet been established and K_s
is usually taken as 1.0.

Table 3.1 Allowable Bending Stress S_{at} for Gear Steels

Material hardness	S_{at} (psi)
180 Bhn	25,000–33,000
300 Bhn	36,000–47,000
400 Bhn	42,000–56,000
Carburized R_c 55	55,000–65,000
Carburized R_c 60	55,000–70,000
Nitrided R_c 60	38,000–48,000

Source: Ref. 4.

K_m = load distribution factor. The load distribution factor accounts for inaccuracies in the bearing bore locations leading to misalignment of the axes of rotation, alignment errors due to gear tooth inaccuracies, and deflections due to load or thermal distortion. For face widths less than 2.0 in., accurate gears and mountings and stiff housings a K_m as low as 1.1 may be used. In cases where poor alignment is anticipated K_m may equal 2.0 or more.

The relation of calculated bending stress to the allowable stress of the material is [4]

$$S_t \leq \frac{S_{at}K_l}{K_t K_r}$$

where

S_{at} = allowable material stress, psi (see Table 3.1)
K_l = life factor

Table 3.2 Life Factor—K_l

Number of cycles	160 Bhn	250 Bhn	400 Bhn	Case carb.
Up to 1000	1.6	2.4	3.4	2.7
10,000	1.4	1.9	2.4	2.1
100,000	1.25	1.5	1.7	1.6
1 million	1.1	1.1	1.2	1.2
10 million	1.0	1.0	1.0	1.0

Source: Ref. 4.

The life factor adjusts the allowable stress for the required number of operation. Table 3.2 defines values of K_l. Beyond 10^7 cycles the allowable stress may have to be further derated. A value of K_l of 0.8 might be chosen for 10^{10} cycles.

K_t = temperature factor. The temperature factor is usually taken as 1.0 unless the oil or gear blank temperature exceeds 250°F.

K_r = reliability factor. The allowable bending stress values in Table 3.1 reflect a failure probability of fewer than 1 in 100 in 10^7 load cycles. If a lower statistical probability of failure is desired, K_r must be greater than 1.0. A K_r of 1.25 may reflect a failure probability of 1 in a 1000 and a K_r of 1.5 may reflect a failure probability of 1 in 10000.

Power Rating

Quite often, gears are rated on the basis of power. Equation (3.1) can be manipulated to the following:

$$P_{at} = \frac{n_p \cdot PD \cdot K_v}{126,000 K_a} \quad \frac{F}{K_m} \quad \frac{J}{K_s P_d} \quad \frac{S_{at} K_l}{K_r K_t} \tag{3.2}$$

where

P_{at} = allowable transmitted power on the basis of bending strength, hp
n_p = pinion speed (high-speed member), rpm
PD = pinion pitch diameter, in.

Overloads

When a gear is subjected to infrequent momentary high overloads, the transient stress should be compared to the yield strength of the material. A factor of safety K_r, ranging from 1.33 for conventional industrial gearing to 3.0 for very high reliability, should be applied. Examples of infrequent overloads are equipment that experiences high starting loads once a day or construction machinery that occasionally stalls, incurring high power drain at low speed.

Reverse Bending

When gear teeth experience reverse bending (loading in both directions), such as in an idler or planet gear, 70% of the allowable fatigue strength should be used.

DURABILITY RATING

The durability rating of gear teeth concerns itself with fatigue pitting resistance. Equations based on work by Herz are used to calculate contact (compressive) stresses between the mating gear teeth (Figure 3.6). Gear tooth contact conditions are similar to those between two cylinders except that on gear teeth the radii of curvature are continuously changing. A specific mesh point, such as at the pitch diameters, is chosen at which to make the calculation. Although it is the surface stress which is calculated, subsurface shear stresses proportional to the surface compressive stress are the actual cause of crack initiation. The fundamental equation for compressive stress in a gear mesh is [4]

$$S_c = C_p \sqrt{\frac{W_T}{PD \cdot F} \quad \frac{1.0}{I}} \quad \text{psi}$$

$$C_p = \text{elastic coefficient} = \sqrt{\frac{1.0}{\pi\left[\left(\dfrac{1 - \mu_p^2}{E_p}\right) + \left(\dfrac{1 - \mu_g^2}{E_g}\right)\right]}}$$

where

E_p = pinion modulus of elasticity, psi
E_g = gear modulus of elasticity, psi
μ_p = pinion Poisson's ratio
μ_g = gear Poisson's ratio

For steel-on-steel gears the elastic coefficient C_p = 2300 psi.

I = geometry factor

The geometry factor deals with the radii of curvature at the point of contact and the load sharing between teeth:

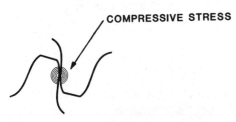

COMPRESSIVE STRESS

Figure 3.6 Compressive stress criterion for durability rating.

$$I = \frac{C_c}{M_n}$$

where

C_c = curvature factor
M_n = load sharing ratio

For helical gears the compressive stress is usually calculated at the pitch line, where

$$C_c = \frac{\cos \phi_t \sin \phi_t}{2.0} \left(\frac{M_g}{M_g \pm 1.0} \right) \quad \text{+ for external gear mesh, - for internal gear mesh}$$

where

ϕ_t = transverse pressure angle, deg
M_g = gear ratio (number of gear teeth divided by number of pinion teeth, always greater than 1.0)

For spur gears the compressive stress is usually calculated at the start of single tooth contact. The curvature factor C_c must be multiplied by a modification factor C_x:

$$C_x = \frac{R_1 R_2}{R_P R_G}$$

where

R_1 = radius of curvature at the lowest point of single tooth contact on the pinion, in.
R_2 = radius of curvature at the highest point of single tooth contact on the gear, in.
R_P = radius of curvature at the pinion pitch diameter, in.
R_G = radius of curvature at the gear pitch diameter, in.

The method of calculation of these radii of curvature was discussed in Chapter 2.

M_n = load-sharing ratio, which depends on the profile and face contact ratios
M_n = 1.0 for spur gears
M_n = F/L_{min} for helical gears
L_{min} = total length of lines of contact (minimum), in.

A procedure for calculating L_{min} was presented in Chapter 2. An estimate for L_{min} which holds for most helical gears when the face contact ratio exceeds 2.0, or when the face contact ratio or transverse contact ratio is an integer of 1.0 or greater, is

$$M_n = \frac{P_n}{0.95Z}$$

where

P_n = normal base pitch, in.
Z = length of action in the transverse plane, in. (see Figure 2.15)

COMPRESSIVE STRESS RATING

In the AGMA rating system the basic compressive stress is modified by several factors that deal with characteristics of a specific application:

$$S_c = C_p \sqrt{\frac{W_t C_a}{C_v} \frac{C_s}{PD \cdot F} \frac{C_m C_f}{I}} \quad \text{psi} \tag{3.3}$$

where C_f is the surface condition factor. This factor depends on the tooth finish, residual stresses, and work-hardening effects. For good-quality gearing it is taken as 1.0. The C_a, C_v, C_s, and C_m factors have the same values as the corresponding K_a, K_v, K_s, and K_m factors discussed previously.

The relation of calculated compressive stress to the allowable stress of the material is

$$S_c \leq S_{ac} \frac{C_l C_h}{C_t C_r}$$

where

S_{ac} = allowable contact stress, psi (see Table 3.3)
C_h = hardness ratio factor. When the pinion is significantly harder than the gear, there is a work-hardening effect on the gear and the gear

Table 3.3 Allowable Contact Stress S_{ac} for Gear Steels

Material hardness	S_{ac} (psi)
180 Bhn	85,000–95,000
300 Bhn	120,000–135,000
400 Bhn	155,000–170,000
Carburized R_c 55	180,000–200,000
Carburized R_c 60	200,000–225,000
Nitrided R_c 60	192,000–216,000

Source: Ref. 4.

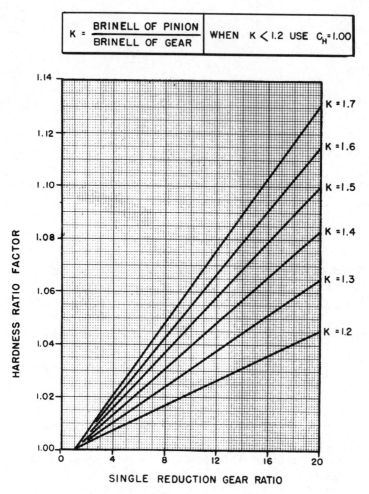

$$K = \frac{\text{BRINELL OF PINION}}{\text{BRINELL OF GEAR}} \quad \text{WHEN } K < 1.2 \text{ USE } C_H = 1.00$$

Figure 3.7   Hardness ratio factor. (From Ref. 4.)

        durability rating is increased. This effect increases with the reduc-
tion ratio as shown in Figure 3.7.

C_1 = life factor. The life factor adjusts the allowable stress for the required number of cycles of operation. A factor of 1.0 is used for
10^7 cycles. For 10^4 cycles and less a factor of 1.5 can be used.
Insufficient data are available to define C_1 beyond 10^7 cycles;
however, it appears there is no well-defined endurance limit and
allowable stress should be decreased as load cycles accumulate over
10^7. A life factor of 0.7 might be used for 10^{10} cycles.

The temperature factor C_t and the reliability factor C_r have the same values as the corresponding K_t and K_r factors discussed previously.

Power Rating

Quite often, gears are rated on the basis of power. Equation (3.3) can be manipulated to the following:

$$P_{ac} = \frac{n_p F}{126,000} \frac{IC_v}{C_s C_m C_f C_a} \left(\frac{S_{ac} \cdot PD}{C_p} \frac{C_l C_h}{C_t C_r} \right)^2 \quad hp \qquad (3.4)$$

where P_{ac} is the allowable transmitted horsepower on the basis of compressive stress.

AGMA STANDARDS FOR ENCLOSED DRIVE RATINGS

AGMA Standard 420.04 [1] covers enclosed drives with pitch line velocities not exceeding 5000 fpm and pinion speeds not exceeding 3600 rpm. Higher-speed enclosed drives are covered by AGMA Standard 421.06 [2], which deals with helical and herringbone gear units. These Standards basically use Eqs. (3.2) and (3.4) for strength and durability rating but simplify the many modifying factors and arrive at a service factor based on transmitted power rather than a reliability factor based on stress.

For the strength rating Standard 420.04 states that

$$P_{at} = \frac{K_1 K_2 K_3 J}{P_d}$$

$$K_1 = \frac{n_p \cdot PD \cdot K_v}{126,000}$$

$$K_2 = \frac{F}{K_m}$$

$$K_3 = S_{at} K_1$$

where

P_{at} = horsepower rating based on tooth strength
J = geometry factor (Figures 3.4a, 3.4b, and 3.4c)
P_d = diametral pitch—transverse, in.$^{-1}$
n_p = pinion rpm
PD = pinion pitch diameter, in.
K_v = $50/(50 + \sqrt{PLV})$ for spur gears

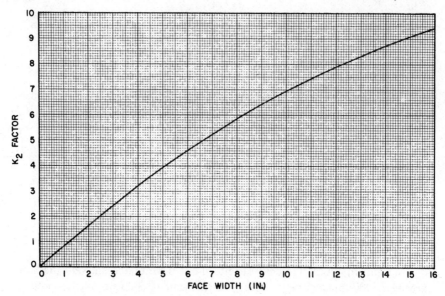

Figure 3.8 K_2 factor for strength rating of spur and helical gearings. (From Ref. 1.)

K_v = $\sqrt{78/(78 + \sqrt{PLV})}$ for helical gears

PLV = pitch line velocity, fpm

K_2 = face width alignment factor as specified in Figure 3.8

K_3 = allowable fatigue stress times life factor as specified in Figures 3.9 and 3.10

For the durability rating Standard 420.04 states:

$$P_{ac} = C_1 C_2 C_3 C_4 \qquad\qquad (3.5)$$

$$C_1 = \frac{n_p \cdot PD^2 \cdot C_v}{126,000}$$

$$C_2 = \frac{F}{C_m}$$

$$C_3 = I(S_{ac}/C_p)^2$$

$$C_4 = (C_1)^2$$

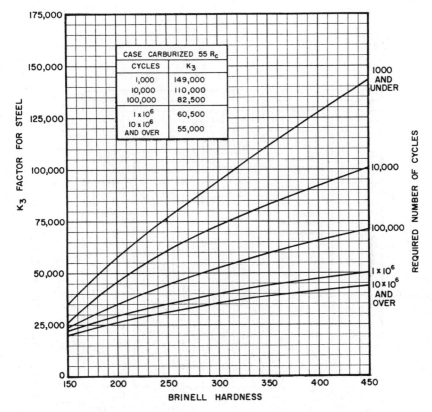

Figure 3.9 K_3 factor for strength rating of spur gears. (From Ref. 1.)

where

P_{ac}	= horsepower rating based on tooth durability
n_p	= pinion rpm
PD	= pinion pitch diameter, in.
C_v	= $50/(50 + \sqrt{PLV})$ for spur gears
C_v	= $78/(78 + \sqrt{PLV})$ for helical gears
PLV	= pitch line velocity, fpm
C_2	= face width/alignment factor as specified in Figure 3.11
I	= geometry factor (see also "Durability" section)
$(S_{ac}/C_p)^2$	= allowable contact stress/elastic coefficient as specified in Figure 3.12
C_4	= life factor[2] as defined in Figure 3.13

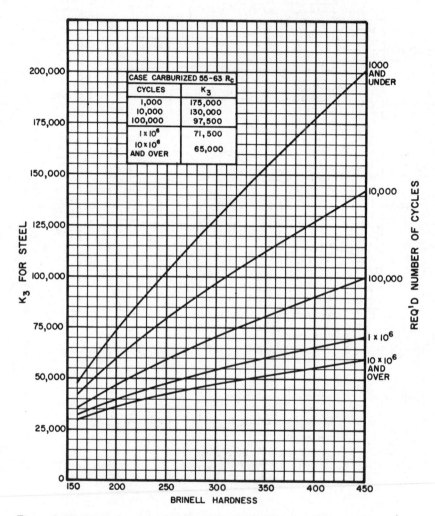

Figure 3.10 K_3 factor for strength rating of helical gears. (From Ref. 1.)

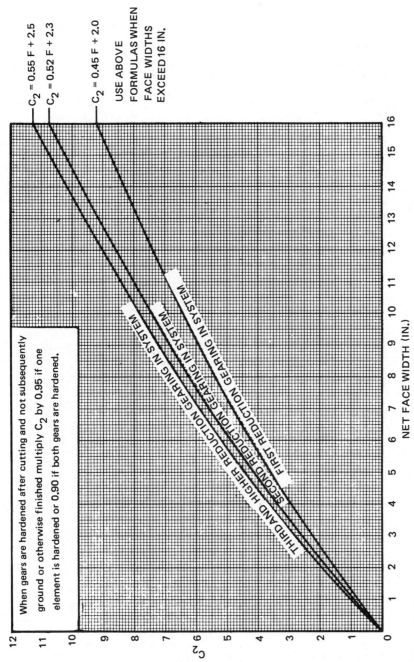

Figure 3.11 C_2 factor for durability rating. (From Ref. 1.)

When gears are hardened after cutting and not subsequently ground or otherwise finished multiply C_2 by 0.95 if one element is hardened or 0.90 if both gears are hardened.

$C_2 = 0.55 F + 2.5$

$C_2 = 0.52 F + 2.3$

$C_2 = 0.45 F + 2.0$

USE ABOVE FORMULAS WHEN FACE WIDTHS EXCEED 16 IN.

NET FACE WIDTH (IN.)

THIRD AND HIGHER REDUCTION GEARING IN SYSTEM

SECOND REDUCTION GEARING IN SYSTEM

FIRST REDUCTION GEARING IN SYSTEM

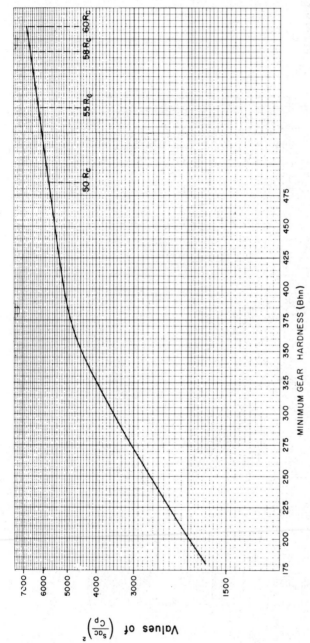

VALUES ARE TO BE TAKEN FROM THE ABOVE CURVE FOR THE MINIMUM HARDNESS SPECIFIED FOR THE GEAR.
VALUES FOR SUGGESTED GEAR AND PINION HARDNESS COMBINATIONS ARE TABULATED BELOW FOR CONVENIENCE

MINIMUM BRINELL HARDNESS

GEAR	180	210	225	245	255	270	285	300	335	350	375	55 R_C	58 R_C
PINION	210	245	265	285	295	310	325	340	375	390	415	55 R_C	58 R_C
$\left(\dfrac{s_{ac}}{C_p}\right)^2$	1750	2100	2300	2560	2700	2950	3200	3460	4200	4460	4820	6200	6600

Figure 3.12 C_3 factor for durability rating. (From Ref. 1.)

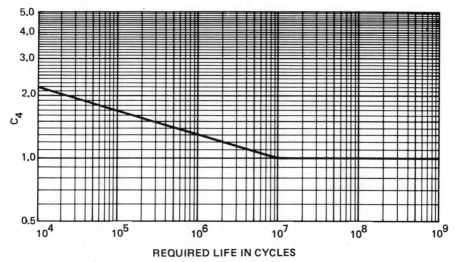

Figure 3.13 C_4 factor for durability rating. (From Ref. 1.)

HIGH-SPEED GEARING

Strength

The high-speed standard, Standard 421.06, uses Eqs. (3.2) as it stands except that the dynamic factor K_v is defined as follows:

$K_v = \sqrt{78/(78 + \sqrt{PLV})}$ for pitch line velocities below 7000 fpm

$K_v = 0.695$ for pitch line velocities 7000 fpm and higher

Durability

The 421.06 Standard uses Eq. (3.5):

$$P_{ac} = C_1 C_2 C_3 C_4$$

with three modifications:

1. $C_4 = 1.0$.
2. The dynamic factor C_v is 0.48 for pitch line velocities of 7000 fpm and higher.
3. The C_2 factor is defined in Figure 3.14.

Service Factors

When P_{at} and P_{ac} are calculated, the rating of the gearbox will be based on the smaller of the strength and durability ratings. A service factor must now be

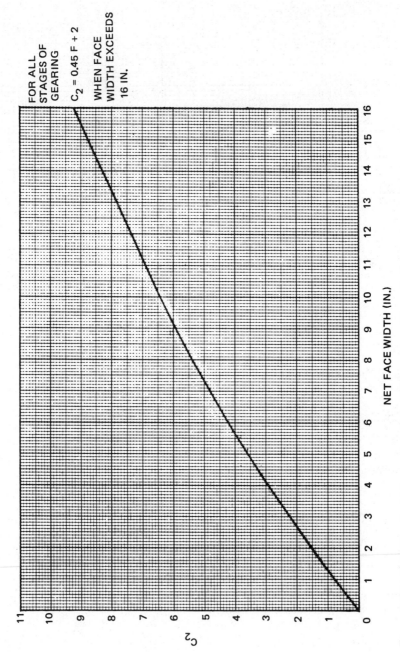

FOR ALL
STAGES OF
GEARING

$C_2 = 0.45 F + 2$

WHEN FACE
WIDTH EXCEEDS
16 IN.

NET FACE WIDTH (IN.)

C_2

Figure 3.14 C_2 factor for high-speed-unit durability rating. (From Ref. 2.)

Table 3.4 Service Factors for High-Speed Units

Application	Service factor		
	Prime mover		
	Motor	Turbine	Internal combustion engine (multicylinder)
Blowers			
Centrifugal	1.4	1.6	1.7
Lobe	1.7	1.7	2.0
Compressors			
Centrifugal—process gas except air conditioning	1.3	1.5	1.6
Centrifugal—air conditioning service	1.2	1.4	1.5
Centrifugal—air or pipeline service	1.4	1.6	1.7
Rotary—axial flow, all types	1.4	1.6	1.7
Rotary—liquid piston (Nash)	1.7	1.7	2.0
Rotary—lobe-radial flow	1.7	1.7	2.0
Reciprocating—3 or more cylinders	1.7	1.7	2.0
Reciprocating—2 cylinders	2.0	2.0	2.3
Dynamometer—test stand	1.1	1.1	1.3
Fans			
Centrifugal	1.4	1.6	1.7
Forced draft	1.4	1.6	1.7
Induced draft	1.7	2.0	2.2
Industrial and mine (large with frequent start cycles)	1.7	2.0	2.2
Generators and Exciters			
Base load or continuous	1.1	1.1	1.3
Peak duty cycle	1.3	1.3	1.7
Pumps			
Centrifugal (all service except as listed below)	1.3	1.5	1.7
Centrifugal—boiler feed	1.7	2.0	
Centrifugal—descaling (with surge tank)	2.0	2.0	
Centrifugal—hot oil	1.5	1.7	
Centrifugal—pipeline	1.5	1.7	2.0
Centrifugal—water works	1.5	1.7	2.0
Dredge	2.0	2.4	2.5
Rotary—axial flow, all types	1.5	1.5	1.8
Rotary—gear	1.5	1.5	1.8
Rotary—liquid piston	1.7	1.7	2.0
Rotary—lobe	1.7	1.7	2.0

Table 3.4 (Continued)

		Service factor	
		Prime mover		
Application	Motor	Turbine	Internal combustion engine (multicylinder)
Rotary–lobe	1.7	1.7	2.0
Rotary–sliding vane	1.5	1.5	1.8
Reciprocating–3 cylinders or more	1.7	1.7	2.0
Reciprocating–2 cylinders	2.0	2.0	2.3
Paper industry			
Jordan or refiner	1.5	1.5	
Paper machine–line shaft	1.3	1.3	
Paper machine–sectional drive	1.5		
Pulp beater	1.5		
Sugar industry			
Cane knife	1.5	1.5	1.8
Centrifugal	1.5	1.7	2.0
Mill	1.7	1.7	2.0

Source: Ref. 2.

assigned to the application in order to determine the horsepower that the unit will be rated against. The specified or actual horsepower transmitted through the gearbox is multiplied by the service factor and called the equivalent or service horsepower. The smaller of P_{ac} and P_{at} must equal or exceed the equivalent horsepower:

$$\text{Service factor} = \frac{\text{smaller of } P_{ac} \text{ or } P_{at}}{\text{specified horsepower}}$$

The AGMA and API Standards [1–3] contain tables of recommended service factors for various applications. Table 3.4 illustrates the numerical values of service factors for high-speed units. The API Standard has very similar values. Lower-speed gearboxes covered by AGMA Standard 420.04 have service factors as shown in Table 3.5, which takes into account the duration of service, prime mover characteristics, and driven machine characteristics.

The AGMA and API standards are good basic guidelines which, when followed, result in gearboxes that perform successfully. They provide a baseline with which to judge competing designs. The enclosed drive standards [1–3] will give more conservative results than the general standard [4], but this is to be

Table 3.5 Service Factors for Low-Speed Units

| | | Driven machine load classification | | |
| | | Uniform | Moderate shock | Heavy shock |
Prime mover	Duration of service			
Electric motor,	Occasional ½ hr per day	0.50	0.80	1.25
steam turbine,	Intermittent 3 hr per day	0.80	1.00	1.50
or hydraulic	Over 3 hr up to and incl.	1.00	1.25	1.75
motor	10 hr per day			
	Over 10 hr per day	1.25	1.50	2.00
Multicylinder	Occasional ½ hr per day	0.80	1.00	1.50
internal	Intermittent 3 hr per day	1.00	1.25	1.75
combustion	Over 3 hr up to and incl.	1.25	1.50	2.00
engine	10 hr per day			
	Over 10 hr per day	1.50	1.75	2.25
Single cylinder	Occasional ½ hr per day	1.00	1.25	1.75
internal	Intermittent 3 hr per day	1.25	1.50	2.00
combustion	Over 3 hr up to and incl.	1.50	1.75	2.25
engine	10 hr per day			
	Over 10 hr per day	1.75	2.00	2.50

Source: Ref. 1.

expected since the enclosed drive standards reflect gear manufacturer's field experience and are based on practical as well as theoretical considerations.

A drawback of the enclosed drive standards is that they do not quantify the effects of component quality or metallurgy in the rating procedures. A given gearbox may contain gears with better metallurgical characteristics or closer tolerances than another, yet both units could claim the same rating according to existing standards. There is work proceeding to improve the standards. If an application is considered critical, the user should determine the specific quality and metallurgical characteristics required and identify them in the procurement document rather than simply call for a unit complying with a particular standard.

Very high speed gearing operating above 20,000 fpm pitch line velocity requires analysis beyond that presented in industry standards. If a procedure such as that outlined in the API standard is used to size a very high speed unit, the gear diameters that result may become so large that centrifugal effects may cause stresses and deflections that override the conventional strength and durability considerations. In some high-speed applications it is preferable to minimize pitch line velocity at the expense of higher bending and compressive stresses. These types of units must be carefully designed and incorporate the highest-quality components with the best metallurgy available.

API STANDARD

The American Petroleum Institute standard [3], like the AGMA standards, is widely used to procure gear units. There are two editions, the first issued in 1968 [8] and the second in 1977 [3]. The first edition rating method conforms to the AGMA high-speed standard. The second edition rating method is more conservative. Gear units are sized on the basis of a tooth pitting index called the K factor. K-factor ratings are common in the industry and often used as a simple rating index:

$$K = \frac{W_t}{F \cdot PD} \left(\frac{M_g \pm 1}{M_g} \right) \quad \text{+ for external gear meshes, - for internal gear meshes}$$

where

K = tooth pitting index
W_t = transmitted tangential load, lb
F = net face, width, in.
PD = pinion pitch diameter, in.
M_g = gear ratio

Table 3.6 Material Index Numbers and Maximum L/d Values[a]

Gear hardness minimum	Pinion hardness minimum	Material index number	Maximum pinion L/d	
			Double-helical	Single-helical
223 Bhn	262 Bhn	130	2.4	1.7
262 Bhn	302 Bhn	160	2.3	1.6
302 Bhn	341 Bhn	200	2.2	1.5
352 Bhn	50 R_c (nitrided)	260	2.0	1.45
50 R_c (nitrided)	50 R_c (nitrided)	300	1.9	1.4
55 R_c (carburized)	55 R_c (carburized)	410	1.7	1.35
58 R_c (carburized)	58 R_c (carburized)	440	1.6	1.3

[a]L, net face width plus gap;
d, pinion pitch diameter.
Source: Ref. 3.

The allowable K is the material index number shown in Table 3.6, divided by the service factor:

$$K = \frac{\text{material index number}}{\text{service factor}}$$

Also shown in Table 3.6 are maximum allowable face width/pinion pitch diameter ratios. In the case of double helical gears the total face width L is the net face width plus the gap between the helices. The bending stress is calculated as follows:

$$S_t = \frac{W_t P_{nd}}{F} \ (SF) \ \frac{1.8 \cos \psi}{J} \quad \text{psi}$$

where

S_t = bending stress, psi
P_{nd} = normal diametral pitch, in.$^{-1}$
F = net face width, in.
SF = service factor
ψ = helix angle, deg
J = geometry factor

The bending stress must not exceed the values shown in Figure 3.15.

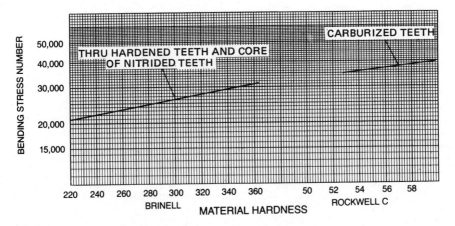

Figure 3.15 Bending stress allowables. (From Ref. 3.)

SAMPLE RATING CASE

Let us look at a specific application and see how the various rating procedures compare. Consider the first mesh of a two-stage high-speed reducer connecting a gas turbine with a generator having the following characteristics:

hp = 4500 hp

Pinion rpm = 14,500 rpm

Pinion pitch diameter = 4.625 in.

Number of pinion teeth = 37

Gear pitch diameter = 14.875 in.

Number of gear teeth = 119

Face width = 6 in.

Transverse pressure angle = 20.3439°

Helix angle = 11.0°

Transverse diametral pitch = 8.0

The gears are single helical, carburized, hardened to R_c 60 and ground to AGMA Quality Class 12. The tangential load is

$$W_T = \frac{63,025(4500)}{14,500} \times \frac{2}{4.625} = 8458 \text{ lb}$$

The pitch line velocity is

$$PLV = \frac{14,500(4.625\,\pi)}{12} = 17,557 \text{ fpm}$$

Using Eq. (3.1), the bending stress is calculated:

K_a = 1.0; smooth operation is anticipated

K_v = 0.94; extrapolated from Figure 3.5

K_s = 1.0

K_m = 1.32, assuming accurate gears and rigid mounting

J = 0.54 for the pinion ⎫
J = 0.56 for the gear   ⎬   Figures 3.4b and 3.4c

$$S_t = \frac{8458(1.0)}{0.94} \times \frac{8}{6} \times \frac{1.0(1.32)}{0.54} = 29,326 \text{ psi}$$

Choosing a life factor $K_l = 0.8$, a temperature factor $K_t = 1.0$, and an allowable material stress of 70,000 psi (Table 3.1), we have

$$K_r = \frac{70,000(0.8)}{(1.0)(29,326)} = 1.91$$

which is a very low probability of failure.

If a reliability factor of 1.0 (a failure probability of 1 in 100) is chosen, the allowable transmitted power on the basis of bending strength from Eq. (3.2) is

$$P_{at} = \frac{14,500(4.625)(0.94)}{126,000(1.0)} \times \frac{6}{1.32} \times \frac{0.54}{1.0(8)} \times \frac{70,000(0.8)}{1.0(1.0)}$$

$$= 8596 \text{ hp}$$

To calculate the compressive stress, use Eq. (3.3):

$$I = \frac{C_c}{M_n}$$

$$C_c = \frac{\cos 20.3439° \sin 20.3439°}{2} \times \frac{3.22}{4.22} = 0.1244$$

$$M_n = \frac{F}{L_{min}} = \frac{6}{10.75} = 0.558$$

$$I = 0.2229$$

$$C_a = 1.0$$

$$C_v = 0.94$$

$$C_s = 1.0$$

$$C_m = 1.32$$

$$C_f = 1.0$$

$$S_c = 2300 \sqrt{\frac{8458(1.0)}{0.94} \times \frac{1.0}{4.625(6)} \times \frac{1.32(1.0)}{0.2229}} = 100,800 \text{ psi}$$

Choosing a life factor $C_l = 0.8$, a hardness factor $C_h = 1.0$, a temperature factor $C_t = 1.0$, and an allowable material stress of 225,000 psi (Table 3.3), we have

$$C_r = \frac{225,000(0.8)(1.0)}{100,800(1.0)} = 1.79$$

Since C_r is lower than K_r, this gear mesh must be rated on the basis of durability rather than tooth strength at the 4500-hp level.

If a reliability factor C_r of 1.0 (a failure probability of 1 in 100) is chosen, the allowable transmitted power on the basis of durability from Eq. (3.4) is

$$P_{ac} = \frac{14{,}500(6)}{126{,}000} \times \frac{0.2229(0.94)}{1.0(1.32)(1.0)(1.0)} \times \left[\frac{225{,}000(4.625)}{2300} \right.$$

$$\left. \times \frac{0.8(1.0)}{1.0(1.0)} \right]^2 = 14{,}360 \text{ psi}$$

As the transmitted power increases, the bending stress increases linearly and the compressive stress increases as the square root of the horsepower; therefore, at a power point of approximately 6300 hp, the bending stress becomes critical. With a reliability factor C_r of 1.0 the bending stress is more critical than the compressive stress and the allowable transmitted power is 8596 hp. The definition of C_r as the reliability factor is relatively new. In the past it has been termed the factor of safety.

On the basis of AGMA Standard 421.06[2], the strength rating is

$$P_{at} = \frac{14{,}500(4.625)(0.695)}{126{,}000(1.0)} \times \frac{6}{1.32} \times \frac{0.54}{(1.0)(8)} \times \frac{70{,}000(0.8)}{(1.0)(1.0)}$$

$$= 6355 \text{ hp}$$

The durability rating as per AGMA 421.06 is

$$C_1 = \frac{14{,}500(4.625)^2(0.48)}{126{,}000} = 1.18$$

$$C_2 = 4.2 \quad \text{(Figure 3.14)}$$

$$C_3 = I(S_{ac}/C_p)^2$$

$$I = 0.2229$$

$$(S_{ac}/C_p)^2 = 6800 \quad \text{(Figure 3.12)}$$

$$P_{ac} = 7513 \text{ hp}$$

The service factor at 4500 hp is

$$SF = \frac{6355}{4500} = 1.41$$

On the basis of API Standard 613 [3], the strength rating is

$$S_t = \frac{8458(8.15)}{6} \; (SF) \; \frac{1.8 \cos 11°}{0.54}$$

From Figure 3.15 the allowable bending stress number is 40,000 psi and the service factor is

$$SF = \frac{40,000(6)(0.54)}{8458(8.15)(1.8)(\cos 11)} = 1.06$$

The API durability rating is

$$K = \frac{8458}{6(4.625)} \times \frac{4.22}{3.22} = 399$$

The allowable material index number from Table 3.4 is 440, and in this case the service factor is less than 1.0:

$$SF = \frac{399}{440} = 0.91$$

SCORING

High-speed gearing (above 5000 fpm or 3600 rpm) operating with low-viscosity lubricants is prone to a failure mode called scoring. In contrast to the classic failure modes pitting and breakage, which generally take time to develop, scoring occurs early in the operation of a gear set and can be the limiting factor in the gear's power capability. Scoring failures and the degree of scoring that may be accepted are described in Chapter 12.

FLASH TEMPERATURE INDEX

The critical total temperature hypothesis (flash temperature index) [9] appears to be the most reliable method of analysis presently used to predict scoring. It states that scoring will occur when a critical total temperature, which is characteristic of the particular combination of lubricant and gear material, is reached.

$$T_f = T_b + \Delta T$$

where

T_f = flash temperature index, °F
T_b = gear blank temperature, °F
ΔT = maximum rise of instantaneous surface temperature in the tooth mesh above the gear blank's surface temperature

The gear blank temperature is difficult to estimate. It may be significantly higher than the bulk oil temperature [10]. The heat transfer capability of the gear must be considered in attempting to estimate this parameter. Often, the blank temperature is approximated as the average of the oil temperature entering and leaving the gearbox.

One form of the fundamental flash temperature index formula is [11]

$$T_f = T_b + \frac{C_f f W (V_1 - V_2)}{(\sqrt{V_1} + \sqrt{V_2})\sqrt{B_c/2}}$$

where

C_f = material constant for conductivity, density, and specific heat
f = friction coefficient
V_1 = rolling velocity of pinion at point of contact, fps
V_2 = rolling velocity of gear at point of contact, fps
B_c = width of band of contact
W = specific loading, normal load divided by face width, lb/in.

For steel on steel gears, taking C_f as 0.0528, f as a constant 0.06, and adding a term taking surface finish into account [10], the following equation results [12]:

$$T_f = T_b + \left(\frac{W_{te}}{F_e}\right)^{3/4} \frac{50}{50 - S} Z_t (n_p)^{1/2}$$

where

W_{te} = effective tangential load, lb
F_e = effective face width (use minimum contact length for helical gears), in.
S = surface finish (after running in), rms
n_p = pinion rpm
Z_t = scoring geometry factor

$$Z_t = \frac{0.0175 \left[\sqrt{e_p} - \sqrt{\dfrac{N_p}{N_g}}\, e_g\right]}{(\cos \phi_t)^{3/4} \left[\dfrac{e_p e_g}{(e_p + e_g)}\right]^{1/4}} \qquad Note: \text{ Use absolute value of } Z_t$$

where

e_p = pinion radius of curvature, in.
e_g = gear radius of curvature, in.
N_p = number of pinion teeth (smaller member)

N_g = number of gear teeth (larger member)

ϕ_t = pressure angle, transverse operating

The 50/(50 - s) term was developed by Kelly [10] in an experimental program using gears with surface finish in the range 20 to 32 rms. For gears with surface finish rougher than this range, if the computed value exceeds 3, a factor of 3 should be used. For gears with surface finishes finer than 20, the resulting computed factor may be conservative.

The term W_{te} must be adjusted to allow for the sharing of load by more than one pair of teeth. The following analysis, which modifies the tooth load depending on the position of the gear mesh along the line of action, was developed by Dudley using spur gears of standard proportions [11]. If a more accurate prediction of tooth load sharing is available to the reader, it would be appropriate to use that analysis.

$$W_{te} = KW_t$$

where W_t is the tangential tooth load in pounds.

1. Unmodified tooth profiles

$$K = \frac{1}{3} + \frac{1}{3} \frac{\theta - \theta_{LD}}{\theta_L - \theta_{LD}}$$

$$\theta_{LD} \lesssim \theta < \theta_L$$

$$K = 1.0$$

$$\theta_L \lesssim \theta \lesssim \theta_H$$

$$K = \frac{1}{3} + \frac{1}{3} \frac{\theta_o - \theta}{\theta_o - \theta_H}$$

$$\theta_H < \theta \lesssim \theta_o$$

2. Modified tooth profiles
   a. Pinion driving

$$K = \frac{6}{7} \frac{\theta - \theta_{LD}}{\theta_L - \theta_{LD}}$$

$$\theta_{LD} \lesssim \theta < \theta_L$$

$$K = 1.0$$

$$\theta_L \lesssim \theta \lesssim \theta_H$$

$$K = \frac{1}{7} + \frac{6}{7} \frac{\theta_o - \theta}{\theta_o - \theta_H}$$

$$\theta_H \; < \theta \lesssim \theta_o$$

b. Pinion driven by gear:

$$K = \frac{1}{7} + \frac{6}{7} \; \frac{\theta - \theta_{LD}}{\theta_L - \theta_{LD}}$$

$$\theta_{LD} \lesssim \theta < \theta_L$$

$$K = 1.0$$

$$\theta_L \; \lesssim \theta \lesssim \theta_H$$

$$K = \frac{6}{7} \; \frac{\theta_o - \theta}{\theta_o - \theta_H}$$

$$\theta_H \; < \theta \lesssim \theta_o$$

where

θ = any pinion roll angle, rad
θ_{LD} = roll angle at the pinion limit (form) diameter, rad
θ_L = roll angle at the lowest point of single tooth contact on the pinion, rad
θ_H = roll angle at the highest point of single tooth contact on the pinion, rad
θ_o = roll angle at the pinion outside diameter, rad

A modified tooth profile would be one that has tip and/or flank relief rather than a true involute form.

The flash temperature index should be calculated at five specific points on the line of action and at several additional points of contact. The five specific points are:

1. Outside diameter of pinion
2. Highest point of single tooth contact
3. Pitch point (flash temperature rise will be zero since there is no sliding)
4. Lowest point of single tooth contact
5. Form (contact diameter) of pinion

Figure 3.16 is a typical plot of flash temperature rise along the line of action.

The most convenient way to generate a plot such as shown in Figure 3.16 is by the use of a computer program. By stepping through successive roll angles, the flash temperature index can be calculated at many points. From Figure 3.16 it can be seen that there will be two peaks, one during the arc of approach (pinion form diameter to pitch diameter) and one during the arc of recess (pinion pitch diameter to outside diameter). To achieve the minimum flash

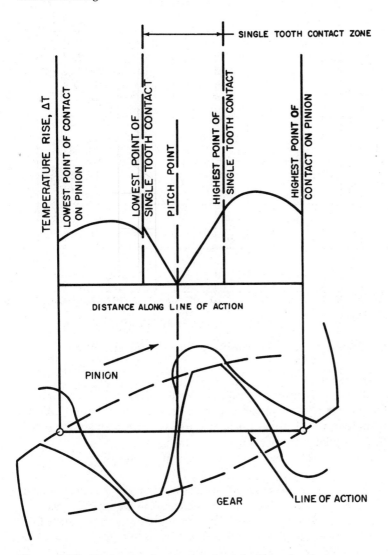

Figure 3.16 Flash temperature rise along the line of action.

temperature index, the flash temperature rise in the arc of approach should be equal to the rise in the arc of recess. An optimum tooth design can be achieved by the use of long and short addendums. The computer program, starting with standard addendums (1/diametral pitch), automatically varies the pinion and gear addendums in defined increments until the optimum flash temperature is

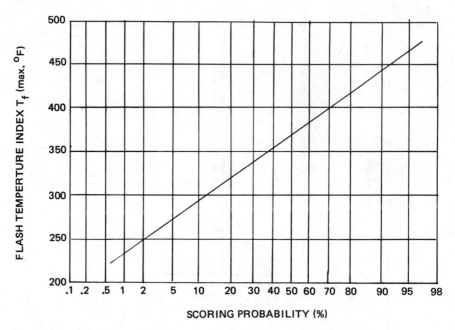

Figure 3.17 Scoring probability versus flash temperature index. (From Ref. 12.)

obtained. With the resulting long and short addendum designs of this nature, standard tooth thicknesses are no longer applicable. If standard tooth thicknesses were utilized, an unbalance of bending stresses between pinion and gear would result. To optimize the bending stresses, the program enters a second iteration procedure, which varies tooth thickness until bending stresses are balanced.

Figure 3.17 presents the results of an aerospace industry survey correlating scoring to flash temperature index. From the data used in the study, it was assumed that a gearset with a calculated index of 276°F or less represents a low scoring risk. A calculated index ranging from 277 to 338°F represents a medium scoring risk where scoring may or may not occur, and a calculated index of 339°F and higher represents a high scoring risk. The data presented in Figure 3.17 reflects cases using Society of Automotive Engineers (SAE) 9310 steel operating with military standard Mil-L-7808 or Mil-L-23699 synthetic oil. The viscosity of these oils is approximately 4 to 6 cSt at 200°F and 18 to 30 cSt at 100°F. Mineral oils such as light turbine oils, used in high-speed industrial applications, are more viscous and therefore may possibly tolerate a higher flash temperature index.

The equation for the flash temperature index assumes a constant coefficient of friction of 0.06. If it is desired to calculate the coefficient of friction at each point on the line of action, the following equation can be used [13]:

$$f = 0.0127 \log_{10} \frac{3.17 \times 10^8}{\mu_0 \, V_s V_t^2 / W}$$

where

f = coefficient of friction
μ_0 = absolute viscosity, cP
V_s = sliding velocity, ips
V_t = sum velocity, ips
W = specific loading, lb/in.

The equation breaks down at the pitch point, where the sliding velocity is 0.0 and the friction coefficient goes to infinity. Using a variable coefficient of friction the flash temperature index formula becomes

$$T_f = T_b + f \left(\frac{W_{te}}{F_e}\right)^{3/4} \frac{50}{50 - S} \left[Z_t (n_p)^{1/2}\right]$$

$$Z_t = \frac{0.2917 \left[\sqrt{e_p} - \sqrt{\dfrac{N_p}{N_g}} \, e_g \right]}{(\cos \theta_t)^{3/4} \left[\dfrac{e_p e_g}{(e_p + e_g)}\right]^{1/4}}$$

SCORING CRITERION NUMBER

A simplified form of the flash temperature index is presented in Ref. 14. A scoring criterion number is defined.

$$\text{Scoring criterion number} = \left(\frac{W_t}{F_e}\right)^{3/4} \frac{n_p^{1/2}}{P_d^{1/4}}$$

where

W_t = tangential driving load, lb
F_e = contacting face width, in.
n_p = pinion rpm
P_d = diametral pitch

Table 3.7 gives scoring criterion numbers for various oils at various gear blank temperatures. If the scoring criterion number is above the values shown in the table, a possibility exists that scoring will be encountered. The gear blank temperature can be taken as the average of the oil-in and oil-out temperatures.

Table 3.7 Critical Scoring Criterion Numbers[a]

Blank temperature (°F):	100	150	200	250	300
Kind of oil	Critical scoring index numbers				
AGMA 1	9,000	6,000	3,000		
AGMA 3	11,000	8,000	5,000	2,000	
AGMA 5	13,000	10,000	7,000	4,000	
AGMA 7	15,000	12,000	9,000	6,000	
AGMA 8A	17,000	14,000	11,000	8,000	
Grade 1065 Mil-0-6082B	15,000	12,000	9,000	6,000	
Grade 1010 Mil-0-6082B	12,000	9,000	6,000	2,000	
Synthetic (Turbo 35)	17,000	14,000	11,000	8,000	5,000
Synthetic (Mil-L-7808D)	15,000	12,000	9,000	6,000	3,000

[a]Scoring number $= \left(\dfrac{W_t}{F_e} \right)^{3/4} \dfrac{n_p^{1/2}}{P_d^{1/4}}$

Source: Ref. 14.

MINIMUM FILM THICKNESS CRITERION

Scoring is a phenomenon that will occur when gears are operating in the boundary lubrication regime [15] rather than with a hydrodynamic or elastohydrodynamic oil film separating the gear teeth. The film thickness can be calculated [16-18] and compared to the combined surface roughness of the contacting elements to determine if metal-to-metal contact is likely to occur. A criterion used to determine the possibility of surface distress is the ratio of film thickness to composite surface roughness:

$$\lambda = \frac{h_{min}}{\sigma}$$

$$\sigma = \sqrt{\sigma_p^2 + \sigma_g^2}$$

where

λ = film parameter
h_{min} = minimum oil film thickness, in.
σ_p = pinion average roughness, rms
σ_g = gear average roughness, rms

The "partial elastohydrodynamic" or "mixed" lubrication regime occurs if the film parameter λ is between approximately 1 and 4. At higher values, full hydro-dynamic lubrication is established and asperity contact is negligible. Below a λ of 1.0 there is a risk of surface distress.

The minimum elastohydrodynamic film thickness is calculated as follows [16,17]:

$$H = \frac{2.65 \, G^{0.54} \, U^{0.70}}{W^{0.13}}$$

$$H = \frac{h_0}{R} \quad \text{(film thickness parameter)}$$

h_0 = minimum film thickness, in.

R = equivalent radius, in.

$$R = \frac{e_p e_g}{e_p \pm e_g} \quad \text{+ external, − internal}$$

e_p = pinion radius of curvature, in.

e_g = gear radius of curvature, in.

$G = \alpha E'$ (materials parameter)

α = pressure viscosity coefficient, in.^{2}/lb (Figure 3.18)

$$\frac{1}{E'} = \frac{1}{2} \left(\frac{1 - \delta_1^2}{E_1} + \frac{1 - \delta_2^2}{E_2} \right)$$

E' = 33×10^6 for steel on steel

δ_1 = pinion Poisson's ratio

δ_2 = gear Poisson's ratio

E_1 = pinion Young's modulus

E_2 = gear Young's modulus

$$W = \frac{W'}{E \, R} \quad \text{(load parameter)}$$

W' = Specific loading, lb/in.

$$U = \frac{\frac{1}{2}(v_1 + v_2)\mu_0}{E' R}$$

$v_1 = w_p e_p$ = rolling velocity of pinion at point of contact, ips

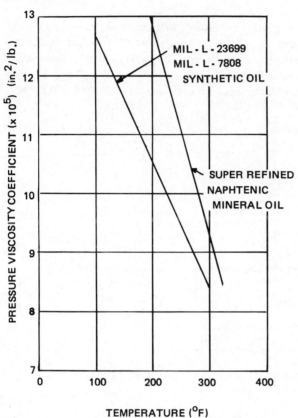

Figure 3.18 Pressure viscosity coefficient versus temperature.

v_2 = $w_g e_g$ = rolling velocity of gear at point of contact, ips

w_p = pinion angular velocity, rad/sec

w_g = gear angular velocity, rad/sec

μ_0 = absolute viscosity, Reyns (lb-sec/in.^{2})

$$\mu_0 = \frac{eZ_k}{6.9(10)^6}$$

e = specific gravity (Figure 3.19)

Z_k = kinematic viscosity, cSt (Figure 3.20)

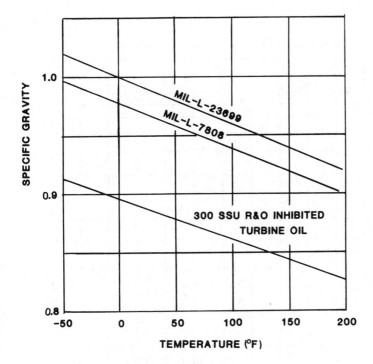

Figure 3.19 Specific gravity versus temperature.

An extensive survey of aerospace power gears operating with synthetic lubricants at high temperature revealed that calculated oil films were on the order of 0.000010 to 0.000020 in. With surface roughness on the order of 20 rms it can be seen that these gears are operating with λ less than 1.0, are in the boundary lubrication regime, and are therefore prone to scoring problems.

Table 3.8 shows the results of a computer analysis of a high-speed gear set with standard addendums. The flash temperature index is the maximum flash temperature rise, 71°F, plus the gear bulk temperature, 160°F. The index of 231°F presents a low scoring risk (Figure 3.17), which could be slightly reduced by optimizing tooth proportions. The calculated coefficient of friction is significantly lower than the assumed valued of 0.06, with a corresponding lower flash temperature rise. The calculated minimum film thickness is 0.000020 with a λ term of 0.71, indicating operation in the boundary lubrication regime.

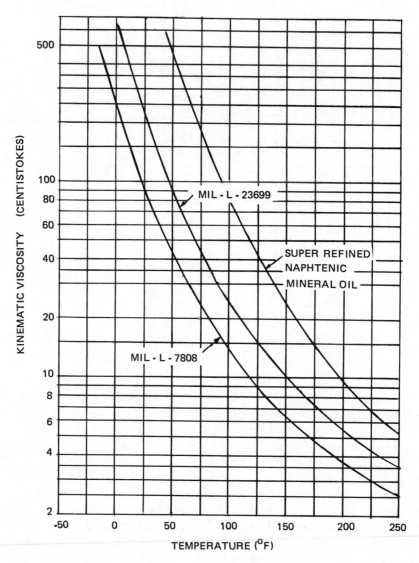

Figure 3.20 Kinematic viscosity versus temperature.

Table 3.8 Scoring Analysis of a High-Speed Gearset

Design parameters

Pinion teeth: 25	Gear teeth: 85
Pinion speed: 12,223 rpm	Horsepower: 281
Face width: 1.0 in.	Helix angle: 0
Diametral pitch: 10	Pressure angle: $25°$
Lubricant type: Mil-L-23699 oil	Gear blank temperature: $160°F$
Oil viscosity: $1.22 (10^{-6})$ lb-sec/in.^{2}	Pressure viscosity coefficient:
Gear material: SAE 9310 steel	$11.4 (10^{-5})$ in.^{2}/lb
	Surface finish: 20 μin. rms

Roll angle, θ (deg)	Friction coefficient, f	Flash temperature rise, ΔT ($°F$)[a]	Flash temperature rise, ΔT ($°F$)[b]	EHD film thickness, h_{min} (μin.)	Film parameter, λ
15.32	0.0	0	0	0	0.0
16.41	0.014	0	40	22	0.77
17.50	0.018	18	58	21	0.73
18.60	0.021	23	68	20	0.72
19.69	0.023	28	71	20	0.71
20.78	0.025	30	70	20	0.71
21.87	0.027	30	64	20	0.72
22.96	0.030	30	60	20	0.71
24.05	0.032	22	42	21	0.73
25.14	0.035	13	25	21	0.75
26.23	0.041	5	7	22	0.76
27.32	0.040	7	9	22	0.78
28.41	0.034	13	25	23	0.80
29.50	0.031	22	41	23	0.81
30.59	0.028	25	52	24	0.84
31.68	0.026	25	59	25	0.88
32.77	0.024	25	62	26	0.91
33.86	0.021	22	61	27	0.96
34.95	0.019	18	57	29	1.01
36.04	0.016	13	48	30	1.08
37.13	0.012	7	33	34	1.19

[a]f is variable.
[b]f held constant at 0.06.
Source: Preventing Gear Tooth Scoring, *Machine Design*, March 20, 1980, Penton/IPC.

SHAFT RATING

Three elements of shaft design are covered in this section:

Shaft stresses
Keyways
Splines

SHAFT STRESSES

Figure 3.21 gives maximum allowable stresses in a shaft due to torsion and bending. These allowable stresses provide for a stress concentration not exceeding 3.0

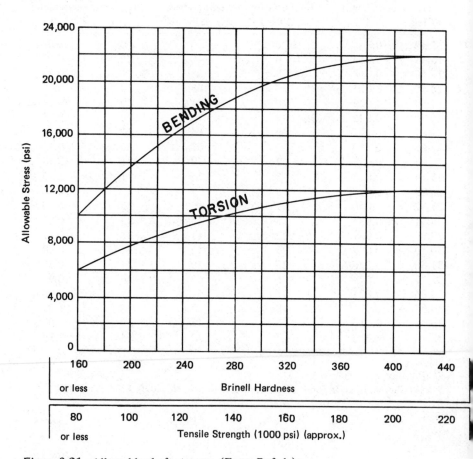

Figure 3.21   Allowable shaft stresses. (From Ref. 1.)

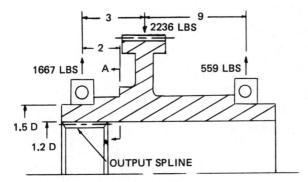

Figure 3.22 Calculation of shaft stress.

and are based on a service factor of 1.0. It is assumed that over 1 million operating cycles will be accumulated. If the shaft is designed to a finite life of less than 1 million cycles, higher allowables can be used.

Let us review a simple example to demonstrate shaft design. The spur gear shaft shown on Figure 3.22 transmits 500 hp at 10,000 rpm. The torque is therefore 3151 in.-lb and if the pitch diameter is 3.0 in., the tangential load is 2101 lb. For a 20° pressure angle the separating load is 765 lb and the resultant radial load is 2236 lb. The bearing reactions are 559 and 1677 lb, respectively.

To find the bending stress at section AA, use

$$S_b = \frac{Mc}{I}$$

where

M = bending moment, in.-lb
c = radius where stress is calculated, in.
I = moment of inertia of section, in.^{4}, for a hollow shaft
   $I = (\pi/64)(\text{O.D.}^4 - \text{I.D.}^4)$

The bending moment at section AA is

$$M = 1677(2.0) = 3354 \text{ in.-lb}$$

$$I = \frac{\pi}{64}(1.5^4 - 1.2^4) = 0.147 \text{ in.}^4$$

and

$$S_b = \frac{3354(0.75)}{0.147} = 17,112 \text{ psi}$$

To find the torsional stress, use

$$S_s = \frac{Tr}{J}$$

where

T = torque, in.-lb
r = radius, in.
J = polar moment of inertia, in.^{4}; for a hollow shaft
      $J = (\pi/32)(\text{O.D.}^4 - \text{I.D.}^4)$

At section AA:

$$J = \frac{\pi}{32}(1.5^4 - 1.2^4) = 0.293 \text{ in.}^4$$

and

$$S_s = \frac{3151(0.75)}{0.293} = 8065 \text{ psi}$$

From Figure 3.21 it can be seen that the shaft hardness must be over 240 Bhn for the stress levels calculated. The analysis given above for shaft bending and torsion will yield satisfactory results for most applications. If it is desired to minimize shaft weight or if a shaft has an unusual configuration, a detailed analysis combining tensile, shear, and compressive stresses must be conducted. Also, stress concentrations must be calculated accurately. When refining a shaft analysis in order to minimize shaft weight, deflections must also be considered, since this factor may be more critical than stress.

KEYWAYS

Keyways must be analyzed for shear and compressive stresses as follows [6]:

$$S_s = \frac{2T}{dwL}$$

$$S_c = \frac{2T}{dh_1 L}$$

where

S_s = shear stress, psi
S_c = compressive stress, psi
T   = shaft torque, lb-in.

Table 3.9 Allowable Keyway Stresses

		Allowable stress (psi)	
Key material	Hardness Bhn	Shear	Comp.
AISI 1018	None specified	7,500	15,000
AISI 1045	225-265	10,000	20,000
	225-300	15,000	30,000
AISI 4140	310-360	20,000	40,000

Source: Ref. 1.

d   = shaft diameter, in. (for taper shaft use mean diameter)
w = width of key, in.
L   = length of key, in.
h_1 = height of key in the shaft or hub that bears against the keyway; for designs where unequal portions of the keyway are in the hub or shaft, h_1 must be the minimum portion

Allowable keyway stresses are presented in Table 3.9.

SPLINE RATING

Standard SAE splines are rated on the basis of tooth shear and compressive stress:

$$S_s = \frac{8T}{D^2 \pi F_e}$$

$$S_c = \frac{2TK_m}{D^2 F_e}$$

where

T   = torque, in.-lb
D = spline pitch diameter, in.
F_e = effective face width, in.
K_m = load distribution factor, Table 3.10

The compressive stress equation simply divides the tangential load per tooth by the tooth bearing area:

$$\text{Tangential load per tooth} = \frac{2T}{D(\text{number of teeth})}$$

Table 3.10 Spline Load Distribution Factor (K_m)

Misalignment (in./in.)	Face width of splines			
	0.5	1.0	2.0	4.0
0.001	1.0	1.0	1.0	1.5
0.002	1.0	1.0	1.5	2.0
0.004	1.0	1.5	2.0	2.5
0.008	1.5	2.0	2.5	3.0

Source: Ref. 19.

$$\text{Bearing area} = F_e(\text{tooth height})$$

$$\text{Tooth height} = 2(\text{spline addendum})$$

$$= \frac{2}{\text{denominator of diametral pitch}}$$

$$= \frac{1}{\text{numerator of diametral pitch}}$$

$$= \frac{\text{pitch diameter}}{\text{number of spline teeth}}$$

For instance, a 20-tooth, 20/40-diametral pitch spline has a tooth height of 0.5 in.

The shear stress equation divides the tangential load by the shear area at the pitch line, which is

$$\text{Shear area} = \frac{F_e \pi}{2(\text{numerator of diametral pitch})}$$

$$= \frac{F_e \pi(\text{pitch diameter})}{2(\text{number of teeth})}$$

In the shear stress equation the assumption is made that only half the spline teeth are in contact. Allowable shear stresses for splines are 50,000 psi for hardened splines (Rc 60) and 40,000 psi for splines in the range Rc 33 to 38.

To determine the compressive stress allowable, the S_c calculated above is modified by the following factors [19]:

$$\text{Sliding splines} - S_c' = \frac{S_c K_a}{L_w}$$

Table 3.11 Spline Application Factor (K_a)

		Type of load		
Power source	Uniform	Light shock	Intermittent shock	Heavy shock
Uniform	1.0	1.2	1.5	1.8
Light shock	1.2	1.5	1.8	2.1
Medium shock	2.0	2.2	2.4	2.8

Source: Ref. 19.

Table 3.12 Spline Wear Factor (L_w)

Number of shaft revolutions	Wear factor, L_w
10,000	4.0
100,000	2.8
1 million	2.0
100 million	1.0
1 billion	0.7
10 billion	0.5

Source: Ref. 19.

Table 3.13 Spline Fatigue Life Factor (L_f)

	Life factor, L_f	
Number of torque cycles	Unidirectional	Fully reversed
10^3	1.8	1.8
10^4	1.0	1.0
10^5	0.5	0.4
10^6	0.4	0.3
10^7	0.3	0.2

Source: Ref. 19.

Table 3.14 Allowable Compressive Stress for Splines

Material	Hardness	Allowable stress (psi)	
		Straight	Crowned
Steel	160–200 Bhn	1,500	6,000
Steel	230–260 Bhn	2,000	8,000
Steel	33-38 R_c	3,000	12,000
Surface hardened	48-53 R_c	4,000	16,000
Case hardened	58-63 R_c	5,000	20,000

Source: Ref. 19.

$$\text{Fixed or locked splines} - S_c' = \frac{S_c K_a}{9 L_f}$$

where

K_a = application factor (Table 3.11)
L_w = wear factor (Table 3.12)
L_f = life factor (Table 3.13)

The life factor is based on the number of times the torque is applied and removed in the expected life of the machine (number of starts). The wear factor is based on the revolutions of the splines. Each revolution there is a back-and-forth rubbing of the teeth, which causes wear. The value of S_c' should not exceed the allowable compressive stress numbers given in Table 3.14.

REFERENCES

1. AGMA Standard 420.04, Practice for Enclosed Speed Reducers or Increasers Using Spur, Helical, Herringbone and Spiral Bevel Gears, American Gear Manufacturers Association, Arlington, Va., December 1975.
2. AGMA Standard 421.06, Practice for High Speed Helical and Herringbone Gear Units, American Gear Manufacturers Association, Arlington, Va., January 1969.
3. API Standard 613, 2nd ed., Special Purpose Gear Units for Refinery Services, American Petroleum Institute, Washington, D.C., February 1977.
4. AGMA Standard 218.01, Rating the Pitting Resistance and Bending Strength of Spur and Helical Involute Gear Teeth, American Gear Manufacturers Association, Arlington, Va., 1982.
5. AGMA Information Sheet 226.01, Geometry Factors for Determining the Strength of Spur, Helical, Herringbone and Bevel Gear Teeth, American Gear Manufacturers Association, Arlington, Va., August 1970.

6. AGMA Gear Handbook 390.03, Gear Classification, Materials and Measuring Methods for Unassembled Gears, American Gear Manufacturers Association, Arlington, Va., January 1973.
7. Buckingham, E., *Analytical Mechanics of Gears*, McGraw-Hill, New York, 1949.
8. API Standard 613, first ed., High-Speed Special-Purpose Gear Units for Refinery Services, American Petroleum Institute, Washington, D.C., August 1968.
9. Blok, H., Surface Temperatures Under Extreme Pressure Conditions, 2nd work, Petroleum Congress, Paris, September 1937.
10. Kelley, B. W., A New Look at the Scoring Phenomena of Gears, SAE National Tractor Meeting, September 1952.
11. Dudley, D. W., *Practical Gear Design*, McGraw-Hill, New York, 1954.
12. AGMA Information Sheet 217.01, Gear Scoring Design Guide for Aerospace Spur and Helical Power Gears, American Gear Manufacturers Association, Arlington, Va., October 1965.
13. Benedict and Kelley, Instantaneous Coefficients of Gear Tooth Friction, *ASLE Transactions 4*, 1961, pp. 59–70.
14. Dudley, D. W., *Gear Handbook*, McGraw-Hill, New York, 1962, pp. 13–44, 13–45.
15. USAAMRDL-TR-75-33, Gear Tooth Scoring Investigation, Southwest Research Institute, San Antonio, Tex., July 1975.
16. Dowson, D., Interdisciplinary Approach to the Lubrication of Concentrated Contacts, NASA SP-237, July 1969.
17. Dowson, D., The Role of Lubrication in Gear Design, British Institution of Mechanical Engineers, 1970.
18. Zaretsky, E. V. and Anderson, W. J., How to Use What We Know About EHD Lubrication, *Machine Design*, November 7, 1968, pp. 167–173.
19. Dudley, D., When Splines Need Stress Control, *Product Engineering*, December 1957.

4
BEARINGS AND SEALS

BEARINGS

One of the first questions to be asked when reviewing a gear box design is: "How are the shafts supported"? Each shaft must be radially and axially located by bearings and proper bearing design and application is at least as important to the operation of the unit as the gears themselves. In fact, bearing problems are more common in gear boxes than are gear failures. A basic choice when designing a unit is which type of bearing to use. There are two general classes, one being journal-type bearings (sometimes called fluid film or slider), the other being rolling-element-type bearings (sometimes called antifriction).

Journal bearings are characterized by operating with a relatively thick oil film between the rotating and stationary elements, the oil film being sheared by relative sliding. Rolling element bearings include ball, roller, and needle bearings with many design variations in each of these classes. The rolling elements are in intimate contact with one another, the oil film being relatively thin compared to journal bearings. In many applications it is not clear which configuration to use. Let us look at various parameters as they relate to journal and rolling element bearings.

1. *Load capacity*. As will be seen later in the chapter, rolling element bearing fatigue life can be calculated on a statistical basis; therefore, in critical applications where load is significant and/or speeds are high, finite lives are predicted. Journal bearings are thought of as having infinite lives provided that the loading is kept below a predetermined level. Therefore, in many applications requiring long life, journal bearings are chosen since the calculated rolling element life is considered insufficient. It is true that a journal bearing when

operating well can last indefinitely while a rolling element bearing eventually is prone to fatigue failure; however, the L_{10} bearing life concept is probably conservative and rolling element bearings quite often operate far longer than predicted lives.

In many applications load is proportional to speed; however, if load is applied at startup, rolling element bearings have an advantage over journal bearings since a journal bearing must attain some speed before developing a hydrodynamic oil film. In some cases journal bearings are externally pressurized at startup so that the load is carried by this pressure rather than a hydrodynamic oil film. This is often done with heavy rotors such as large generator or turbine shafts.

2. *Speed.* Journal and rolling element bearings are both speed limited but for different reasons. As journal bearing speeds increase, the conjunction zone becomes more turbulent and lubricant shearing increases the temperatures, thereby decreasing the oil film thickness. Surface velocities at the bearing bore of 15,000 ft/min are considered high, although operation to 30,000 ft/min has been attained. Rolling element bearing speeds are limited by the centrifugal forces generated between the balls or rollers and outer races, which can exceed the capacity of the bearing. Rolling element bearing speeds are characterized by the DN value (shaft diameter in millimeters times speed in rpm). DN values of over 1×10^6 are considered high. Some gas turbine advanced applications incorporate bearings operating at 3×10^6 DN; however, these bearings require extensive development in such areas as lubrication and retainer design for proper operation.

3. *Lubrication.* Rolling element bearings require significantly less oil flow than journal bearings. As a result, they exhibit less power consumption and heat generation. They also have lower starting torque requirements, particularly at low temperatures where lubricant viscosity is high. Because of their large operating oil films, journal bearings provide some damping in the mechanical system, whereas rolling element bearings are stiff. Journal bearings are very sensitive to dirt in the lubricant since the operating surface is much softer than the hardened contact surfaces of rolling element bearings. Sometimes rolling element bearings are chosen over journal bearings because pre- or postlubrication can be eliminated. It is not necessarily true that journal bearings need to be lubricated prior to startup, and many applications depend on a soft external bearing layer such as babbitt to survive the initial cycles until an oil film forms. For a given application, testing is the only positive way to determine if prelubrication is necessary. Postlubrication is rarely necessary for either rolling element or journal bearings, being used only if some high soak-back temperature will exceed the stabilization temperatures of the bearing materials or cause oil left in the bearing cavity to coke up.

4. *Cost.* In very large quantities journal bearings can be produced at lower cost than rolling element bearings. An example of this is crank shaft bearings in the auto industry. In small quantities for special designs, journal bearings are more expensive. The rolling element bearing industry has standardized on a series of bearing types and manufactures these in large quantities as off-the-shelf items. Equivalent journal bearings for gearboxes, although sometimes listed in catalogs, are not mass produced and in the quantities purchased are relatively expensive.

In many cases the cost of replacing a bearing and the downtime involved is far more significant than the original cost. An example of such a case is a gearbox in a critical process in a petroleum plant. In such an instance journal bearings are preferred, for three reasons:

As discussed before, journal bearings, because the moving parts are separated by a large oil film, can last indefinitely, whereas rolling element bearings are subject to fatigue failures.

The types of failure encountered with journal bearings are less catastrophic and easier to detect. For instance, a journal bearing that is wearing can be monitored with proximity probes and replaced at a convenient time. A spalling rolling element bearing is harder to detect, is introducing abrasive debris into the lubrication system, and may fail abruptly as surface deterioration progresses.

Journal bearings can be split, which enables the user to replace a defective bearing without removing the shafts from the gearbox. This way the couplings remain in place and a time-consuming realignment is not required.

In a broad overview of when journal or rolling element bearings are used: Journal bearings tend to be chosen for high-speed operation, where long life and extreme reliability are desired and the bearing cost is small compared to the cost of system downtime. Lower speed, less critical gearboxes tend to incorporate rolling element bearings. In some sophisticated applications such as aircraft turbine engines, rolling element bearings are used to take advantage of low oil flow requirements, high efficiency, ability to start without oil and at low ambient temperatures.

Once the choice between journal or rolling element bearings has been made, a more detailed analysis of bearing selection and application is in order. The next section will be a review of rolling element bearing fundamentals followed by a consideration of journal bearings.

Rolling Element Bearings

Rolling element bearings are composed of four elements: an inner race, an outer race, a complement of balls or rollers, and a ball or roller separator (sometimes

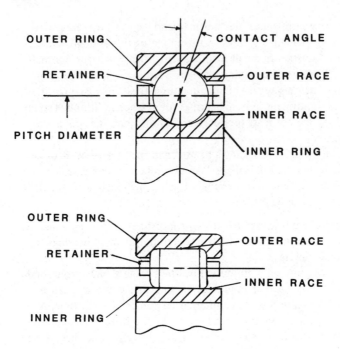

Figure 4.1   Bearing nomenclature.

called a retainer or cage). Figure 4.1 illustrates bearing nomenclature. Usually, the inner race rotates and the outer race is stationary, but there are applications where the reverse is true or both races rotate. Ball bearings in addition to reacting radial loads can carry significant thrust loads, whereas roller bearings though having greater radial load capacity than ball bearings can react only small thrust loads. The greater roller bearing radial capacity is due to the fact that rollers are in line contact with the races as compared to the point contact of ball bearings; therefore, the radial load is spread over a greater area. The following paragraphs describe the most widely used rolling element bearing configurations.

Deep Groove Ball Bearings   This configuration, sometimes called the Conrad type, is illustrated in Figure 4.2. This type of bearing is commonly used in a wide variety of applications where either thrust or radial capacity or both are required. The bearing is assembled by bunching the balls in one arc of the circumference, which permits the inner and outer races to be located in their proper positions. The balls are then separated around the circumference and their retainer fitted in place. A number of retainer configurations are used, the most common being a pressed steel design with the two halves riveted together.

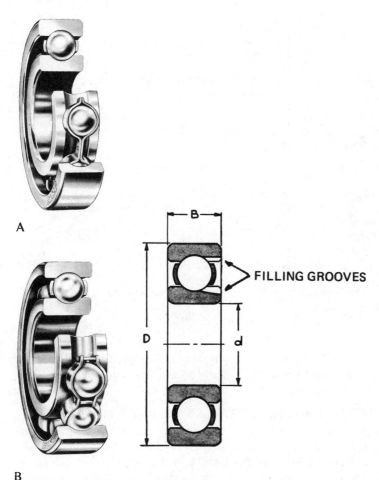

A

B

Figure 4.2 Deep groove ball bearing. A. Conrad type. B. Filling groove type. (Courtesy of SKF Industries, Inc., King of Prussia, Pa.)

Because the load capacity is dependent on the number of balls, several design variations are available which increase the ball complement. One such design is the filling notch bearing, where a single notch is ground into both the inner and outer race through which additional balls can be fed. For maximum strength and high-speed operation a split inner ring bearing can be used where the inner race is made of two sections. In addition to allowing the maximum complement of balls to be assembled, a strong one-piece machined steel retainer can be used in this configuration. It is also possible to split the outer race axially in order to

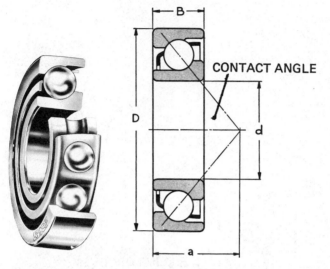

Figure 4.3 Angular contact bearing. (Courtesy of SKF Industries, Inc., King of Prussia, Pa.)

assemble more balls. In the case of the split inner race design the balls do not touch the split, so no reduction in fatigue life occurs. This is not the case with the split outer race bearings.

Angular Contact Ball Bearings Angular contact bearings are designed for applications with high thrust loads in one direction. Figure 4.3 shows the design features, which include one high thrust shoulder on the inner and outer race on opposite sides. There is a low shoulder opposite the high shoulder on each race; thus the maximum number of balls can be snapped into place during assembly. The contact angle between the balls and raceways is designed to be higher than a deep groove bearing giving more thrust capacity. Angular contact bearings can be used in pairs to provide rigid axial location and high thrust capacity in either direction. These bearings can react a combination of thrust and radial loads but are usually used when the thrust load is predominant.

Duplex Ball Bearings Duplex ball bearings are a pair that can be mounted in four ways, as shown in Figure 4.4, to accommodate different loading conditions and stiffness requirements. The inner and outer races are machined such that there is a controlled axial relationship between the two. When the bearings are mounted in the tandem configuration, the precise machining will enable the bearings to share the thrust load equally. In this case the contact angles are parallel. In the face-to-face configuration the contact angle lines converge

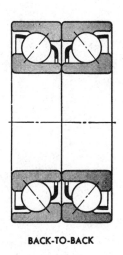

BACK-TO-BACK

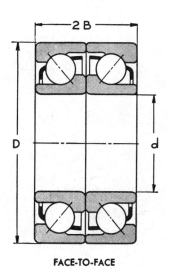

FACE-TO-FACE

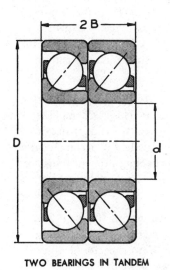

TWO BEARINGS IN TANDEM

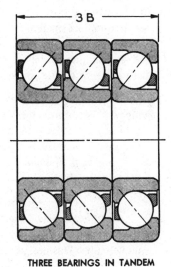

THREE BEARINGS IN TANDEM

Figure 4.4   Duplex bearing arrangements. (Courtesy of SKF Industries, Inc., King of Prussia, Pa.)

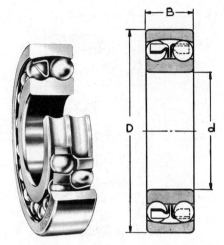

Figure 4.5 Self-aligning ball bearing. (Courtesy of SKF Industries, Inc., King of Prussia, Pa.)

inwardly. This design is used for heavy radial or combined radial and thrust loads or for reversing thrust loads. The face-to-face arrangement allows the bearing to accommodate a small amount of misalignment. The back-to-back configuration has the same large load capacity as the face to face but adds angular rigidity to the system and may be used where it is necessary to restrict misalignment or shaft deflection. The contact angle lines diverge outwardly.

Self-Aligning Ball Bearings Figure 4.5 shows a self-aligning ball bearing which can accommodate large amounts of misalignment. This is accomplished by machining the outer race to a spherical contour; however, the large radius of curvature reduces the bearing-load capacity. Another method of accommodating misalignment is to make the outside diameter of the bearing a spherical surface which then mounts into a spherical seat in the housing. The bearing is then free to position itself.

Double Row Ball Bearings The double row ball bearing (Figure 4.6) consists of one-piece inner and outer races each having two raceways and two ball complements. This design provides heavy radial and thrust capacity in either direction in an envelope somewhat smaller than two single-row bearings. The contact angles can be either face to face or back to back depending on whether it is desired to accommodate misalignment or provide a stiff mounting.

Ball Thrust Bearings Figure 4.7 shows a typical ball thrust bearing, which is similar to a radial ball bearing except that the raceways are axial rather than radial.

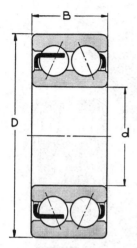

Figure 4.6 Double row ball bearing. (Courtesy of SKF Industries, Inc., King of Prussia, Pa.)

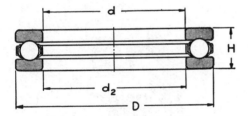

Figure 4.7 Ball thrust bearing. (Courtesy of SKF Industries, Inc., King of Prussia, Pa.)

Single Row

Double Row

Single Row

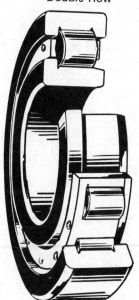

Single Row

Figure 4.8   Cylindrical roller bearings. (Courtesy of SKF Industries, Inc., King of Prussia, Pa.)

Cylindrical Roller Bearings These bearings, illustrated in Figure 4.8, feature cylindrical rollers which run on cylindrical raceways. The roller length is approximately equal to the roller diameter and the rollers are crowned to relieve potentially high stresses at their ends. Roller retainers are positioned either by the rollers or the inner or outer races and are usually two-piece construction either riveted or screwed together. Figure 4.8 shows some of the cylindrical roller bearing configurations available. On the left side are two floating designs where either the inner or outer race has no shoulders and the shaft is allowed to float axially in relation to the housing. Only radial loads can be transmitted. This type of bearing is useful when the shaft must be allowed to move axially due to thermal expansion or deflection due to loading. Cylindrical roller bearings with one shoulder on one race and two shoulders on the other race allow axial movement in one direction and can sustain some thrust load in the other direction. Designs are also available with two shoulders on both the inner and outer races and can sustain axial loads in both directions. In this case one of the races is machined in two pieces for ease of assembly and inspection. It is difficult to determine the amount of thrust load a cylindrical roller bearing is capable of transmitting. The load is carried by sliding contact between the roller

Figure 4.9 Spherical roller bearing. (Courtesy of SKF Industries, Inc., King of Prussia, Pa.)

ends and the race lips such that the action is more like a journal thrust bearing than a rolling element bearing. The important parameters, therefore, are the contact area, surface finish, surface geometry, sliding velocity, lubricant, and operating temperature. In general, a roller bearing can be used to axially locate a shaft when no large thrust loads are anticipated, but should not be expected to react significant amounts of thrust.

Spherical Roller Bearings   Figure 4.9 illustrates a typical spherical roller bearing. There are two rows of rollers which run on a common raceway in the outer ring, which has been ground to a spherical contour on the inner diameter. The inner ring has two raceways ground at an angle to the axis of rotation; thus the bearing is capable of reacting moderate thrust loads in either direction, in addition to heavy radial loads. The spherical outer race enables the bearing to accommodate some shaft misalignment. This type of bearing is suitable for relatively low speed operation.

Tapered Roller Bearings   The tapered roller bearing (Figure 4.10) is designed such that lines extended from each tapered surface intersect at a common point on the bearing axis. Because of the tapered races heavy loads in both the radial and axial directions can be handled. As shown in Figure 4.10, both single row and double row configurations are available. Tapered roller bearings are usually limited to low-speed operation.

Needle Bearings   As shown in Figure 4.11, needle bearings incorporate rollers which are relatively long in relation to their diameter. Their major advantage is high radial load capacity with a small radial envelope requirement. Because of the large length-to-diameter ratio, needle bearings are very sensitive to shaft misalignment.

Bearing Life Rating

Bearings are conventionally rated in terms of L_{10} life, the life that 90% of a group of bearings operating at a given set of conditions will complete or exceed. A bearing failure is defined as the first occurrence of fatigue on one of the rolling elements or on one of the raceways. Fatigue usually manifests itself in the form of a spall. The L_{10} life concept is based on extensive experimental data and as with all fatigue data, there is considerable scatter; however, on the average a plot of life versus percentage of bearings failed takes the form shown in Figure 4.12. In the figure it can be seen that the median life, the life that 50% of a group of bearings will achieve, is five times L_{10} life.

The term "rating life" or just "life" of a bearing has been standardized on as the L_{10} life by organizations such as the Anti-Friction Bearing Manufacturers Association (AFBMA), the American National Standards Institute (ANSI), and the International Standards Organization (ISO). It is a

single row, tapered roller bearings

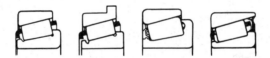

double row, tapered roller bearings

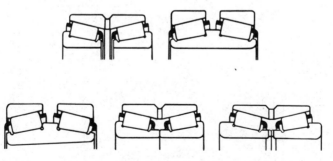

Figure 4.10 Tapered roller bearings. (Courtesy of SKF Industries, Inc., King of Prussia, Pa.)

Figure 4.11 Needle bearing. (Courtesy of SKF Industries, Inc., King of Prussia, Pa.)

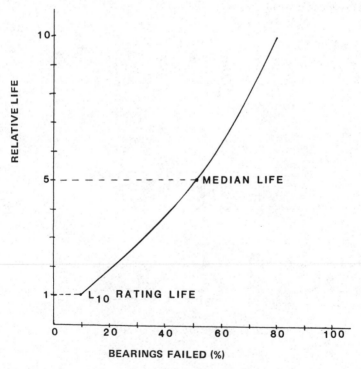

Figure 4.12 Relative bearing life versus percentage failed.

reasonable compromise that provides reliable bearing service while meeting practical economic requirements. In other words, bearings chosen using this criterion will provide reasonably long service, yet not be overly expensive. It must be remembered that the L_{10} life concept pertains only to fatigue of the rolling elements and raceways. Failures due to other causes, such as wear, excessive heat generation, retainer failure, and so on, are not covered by the fatigue model and in many cases limit bearing life.

Extensive testing by the bearing manufacturers, together with analytical studies, has established that the fatigue life of ball bearings is inversely proportional to the third power of the load and the fatigue life of roller bearings is inversely proportional to the 3.33 power of the load:

$$\frac{L_A}{L_B} = \left(\frac{F_B}{F_A}\right)^3 \qquad \text{for ball bearings}$$

$$\frac{L_A}{L_B} = \left(\frac{F_B}{F_A}\right)^{3.33} \qquad \text{for roller bearings}$$

where

L_A = bearing life at condition A, cycles
L_B = bearing life at condition B, cycles
F_A = applied load at condition A, lb
F_B = applied load at condition B, lb

Let us assume the following:

L_B = 1 million cycles
F_B = load (lb) that will give a life of 1 million cycles = C
L_A = L_{10} = rating life, millions of cycles
F_A = P = applied load, lb

From the above are generated the familiar equations

$$L_{10} = \left(\frac{C}{P}\right)^3 \times 10^6 \quad \text{for ball bearings}$$

$$L_{10} = \left(\frac{C}{P}\right)^{3.33} \times 10^6 \quad \text{for roller bearings}$$

The basic load rating C for any bearing can be found tabulated in bearing manufacturers' catalogs. If the L_{10} life is desired in hours:

$$L_{10} = \frac{16,667}{N} \left(\frac{C}{P}\right)^3 \qquad \text{hours for ball bearings}$$

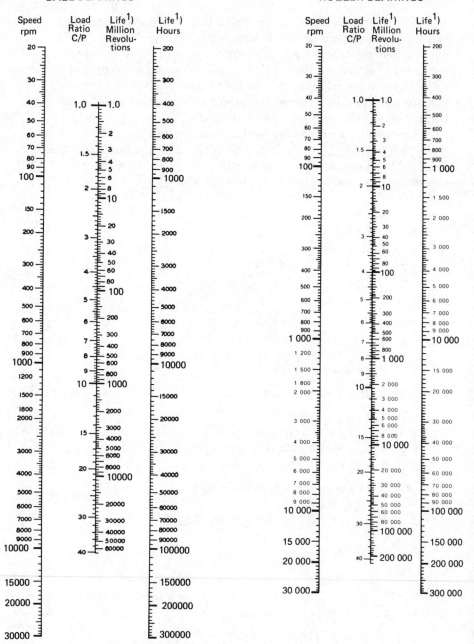

Figure 4.13 Bearing life nomogram. (From Ref. 1.)

$$L_{10} = \frac{16,667}{N} \left(\frac{C}{P} \right)^{3.33} \quad \text{hours for roller bearings}$$

where N is the inner or outer race net speed in rpm. The nomograms in Figure 4.13 represent the equations above and are often used to calculate bearing lives. If the operating load P, operating rpm N, and bearing capacity C are known, a straight line can be drawn that intersects the left scale at the operating speed and the C/P scale at the calculated value of C/P. Where this line intersects the hours scale is the rating life in hours. If the desired rating life, rpm, and applied load are known, a straight line drawn between the speed and life axes will intersect the C/P axis and the required bearing capacity can be calculated.

The L_{10} life concept implies a reliability of 90% with regard to surviving the rating life. In some critical applications reliability numbers greater than 90% are desired. For instance, to achieve a 97% reliability, an L_3 life, the life that 97% of a group of bearings will reach or exceed at a given set of operating conditions, must be calculated. Figure 4.14 gives an estimate of a life adjustment factor corresponding to increased reliability. The factor is based on experimental data and is used as follows:

$$L_n = \alpha_1 L_{10}$$

where

α_1 = life adjustment factor
L_n = fatigue life expectancy for other than 90% reliability

For example, L_3 corresponds to 97% reliability and the factor α_1, from Figure 4.14, is 0.425. If the calculated L_{10} life of a bearing is 10,000 hr, the L_3 life is

$$L_3 = 0.425(10,000) = 4250 \text{ hr}$$

Figure 4.14 indicates that there is no probability of fatigue failure at lives below 5% of the L_{10} rating life. In other words, in the example above the bearing is 100% reliable up to 500 hr of operation.

Rating Life Adjustments

Over the past years substantial increases in the fatigue endurance of rolling element bearings have been made possible by improvements in bearing design, materials, processing, and manufacturing techniques. Also, investigation into effects of high-speed operation and misalignment has generated new insights into their effect on load rating. Using the AFBMA L_{10} life method for determining fatigue life and the bearing load ratings in the various manufacturers' catalogs to calculate the basic L_{10} life, an expected bearing life L_A can be calculated as follows:

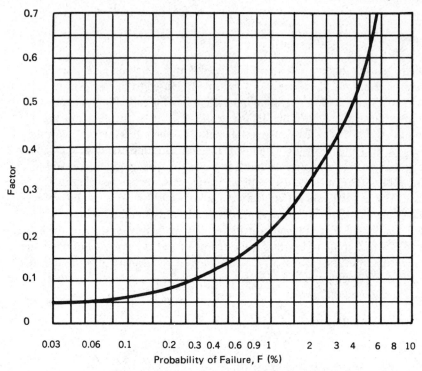

Figure 4.14   Life adjustment factor for reliability. (From Ref. 1.)

$$L_A = (D)(E)(F)(G)(H)L_{10}$$

where D through H are life adjustment factors. This equation, developed by the Rolling Elements Committee of the Lubrication Division of the American Society of Mechanical Engineers and published as Ref. 2, explains the life adjustment factors in detail. The following paragraphs summarize the work.

Material Factor D   The predominant material for rolling element bearings is AISI 52100. AISI is the American Iron and Steel Institute designation for steels of various specific chemistry. The basic bearing dynamic capacities which are presented in catalogs are based on air-melted 52100 steel through-hardened to Rc 58 minimum. The mathematical model to define bearing load capacity evolved in 1949 and since then the bearing steels have been improved such that they are more homogeneous with fewer impurities. A materials factor D of 2 is suggested for currently available steels. Case-hardened materials, used in tapered roller bearings and other applications, have also improved over the years, but insufficient data are available to recommend a materials factor.

Processing Factor E The processing considered here is concerned mainly with the melting practice. Air-melt material is considered the baseline and is assigned an E factor of 1. Vacuum-melted material has an E factor of 3. Although standard catalog bearings are not necessarily vacuum melt, quite often this processing is used and the factor can be taken advantage of.

Lubrication Factor F Bearing life has been found to be significantly affected by the thickness of the lubricant film developed between the contacting elements. Film thickness is affected mainly by speed and lubricant properties at the operating temperature with higher fatigue lives obtained at high speeds or with higher-viscosity lubricants. Conversely, if a bearing operates with poor lubrication film formation due to low speed or insufficient lubricant viscosity, the life predicted from the catalog rating may not be achieved. The oil film between rolling elements of a bearing has been found to be in the elastohydro-dynamic regime and can be calculated according to the methods outlined in Ref. 3 for roller bearings and Ref. 4 for ball bearings. The calculated oil film is compared to the composite roughness of the contacting elements to form what is referred to as the lambda ratio:

$$\lambda = \frac{h}{\sigma}$$

where

λ = lambda ratio
h = oil film thickness, in.
σ = composite surface roughness of the rolling element surfaces, in.

$$\sigma = \sqrt{\sigma_1^2 + \sigma_2^2}$$

where

σ_1 = rms surface finish of body 1, in.
σ_2 = rms surface finish of body 2, in.

The lubrication factor F is proportional to the lambda ratio and varies from approximately 0.2 to 2.8. As an example, let us assume that the balls and race-way of a bearing have rms surface textures of 0.000010 in. The composite surface roughness will be 0.000014 in. If the calculated lubricant film thickness is also 0.000014 in., the lambda ratio is 1. The lubrication factor, F in this case is approximately 1.

Speed Effect Factor G At high speeds the load at the outer race is in-creased because the balls or rollers are subject to centrifugal forces and therefore the fatigue life is reduced. Also, the load distribution on the rolling elements and the zone of loading can change because of centrifugal growth and thermal

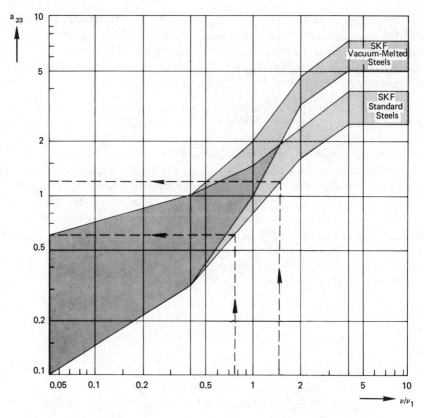

Figure 4.15a Fatigue life adjustment factor a_{23} for bearing materials and lubrication. Lubricant viscosity in application = ν; required lubricant viscosity = ν_1. (From Ref. 5.)

expansion. High-speed bearings must therefore be analyzed using sophisticated computer models which can take all the varying parameters into account, and bearing manufacturers have such programs available for users. In general, centrifugal effects will become significant at DN (bore in millimeters times speed in rpm) values above approximately 0.5×10^6 and certainly bearings operating above 1.0×10^6 DN should be carefully analyzed.

Misalignment Factor H The effect of misalignment on roller bearings is to concentrate the load on one end and therefore reduce the fatigue life. Misalignment can occur due to deflection under load or manufacturing errors in the bearing housing or shaft. The following table provides some misalignment limits beyond which reduction in bearing life may be expected.

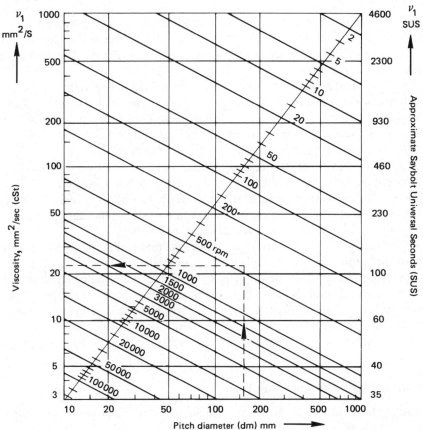

Figure 4.15b Minimum required lubricant viscosity. (Bearing bore + OD) ÷ 2 = dm; required lubricant viscosity for adequate lubrication at the operating temperature = ν_1. (From Ref. 5.)

Allowable Misalignment[a]

	Radians	Minutes
Cylindrical and tapered roller bearings	0.001	3–4
Spherical bearings	0.0087	30
Deep groove ball bearings	0.0035–0.0047	12–16

[a]Based on general experience as expressed in manufacturers' catalogs.
Source: Ref. 2.

When using roller bearings on flexible shafts the misalignment due to load should always be calculated, since this factor may be life limiting.

We can summarize the section on rating life adjustments by noting that today's bearings, assuming good lubrication, can achieve fatigue lives significantly greater than those calculated from catalog ratings. Three- to sixfold life improvement ratings are possible.

One bearing manufacturer's method of presenting these improvements is outlined in [5]. A combined life adjustment factor designated a_{23} replaces the material and lubrication factors described in the preceding section. Figure 4.15a is a plot of the a_{23} factor that can be expected for a particular viscosity ratio (ν/ν_1) and bearing material combination. The viscosity ratio is the ratio of the actual lubricant viscosity (ν), at the operating temperature to the minimum value required (ν_1), which can be found from Figure 4.15b. Figure 4.15b presents the minimum viscosity at the operating temperature required for a given bearing pitch diameter/speed combination. Some common lubricant viscosities at various temperatures can be found in Figures 3.20 and 5.2.

Figure 4.15a is divided into two bands, one for vacuum-melted and one for standard bearing steels. Under comparable operating conditions, certain bearing types (e.g., spherical roller bearings, tapered roller bearings, and spherical roller thrust bearings) normally have a higher operating temperature than other types, such as deep groove ball bearings and cylindrical roller bearings. Therefore, the upper boundary of either band is generally used for ball and cylindrical roller bearings and the lower boundary is used for tapered and spherical roller bearings [5].

Simultaneous Radial and Thrust Loads

Bearings frequently must carry a combination of radial and axial loads. To calculate an L_{10} life, the loading combination must be converted into an equivalent load which would be the constant stationary radial load which if applied to a bearing with rotating inner ring and stationary outer ring would give the same life as that which the bearing would attain under the actual conditions of load and rotation. The general equation for the conversion applying to rotating bearings is

$$P = XF_r + YF_a$$

where

$$
\begin{aligned}
P &= \text{equivalent load, lb} \\
F_r &= \text{actual constant radial load, lb} \\
F_a &= \text{actual constant thrust load, lb} \\
X &= \text{radial factor} \\
Y &= \text{thrust factor}
\end{aligned}
$$

The information required to calculate the X and Y factors for a specific bearing is tabulated in bearing manufacturers' catalogs.

Bearing Selection Example

Figure 4.16 shows a typical page from a bearing catalog. The AFBMA has standardized on bore size, outside diameter, and width of ball and roller bearings and these dimensions are commonly given in millimeters. For instance, in Figure 4.16 a 6020 bearing has a 100-mm bore, a 150-mm outside diameter, and a 24-mm width. This size bearing will be found in any manufacturer's catalog and although the internal geometry will vary, externally all $100 \times 150 \times 24$ mm bearings will be geometrically interchangeable. Let us assume that the 6020 bearing carries a radial load of 1000 lb and a thrust load of 500 lb at a constant inner ring speed of 2500 rpm. Lubrication is oil jet with a 200 SUS viscosity mineral oil. The bearing material is 52100 air melt. The radial and thrust load are combined into an equivalent load using the procedure outlined in Figure 4.16:

$$F_r \quad = 1000 \text{ lb}$$

$$F_a \quad = 500 \text{ lb}$$

$$V \quad = 1.0 \text{ (inner ring rotating)}$$

$$C_o \quad = 9410 \text{ lb (basic static load rating)}$$

$$\frac{F_a}{C_o} \quad = 0.053$$

$$e \quad = 0.26$$

$$Y \quad = 1.71$$

$$\frac{F_a}{V(F_r)} \quad = 0.5 \text{ (larger than e)}$$

$$X \quad = 0.56$$

$$P \quad = 0.56(1)(1000) + 1.71(500) = 1415 \text{ lb}$$

$$C \quad = 10{,}400 \text{ lb (basic dynamic load rating)}$$

$$\frac{C}{P} \quad = 7.35$$

$$L_{10} \quad = \frac{16{,}667}{2500} \left(\frac{C}{P}\right)^3 = 2647 \text{ hr}$$

The bearing pitch diameter is

$$PD = \frac{OD + bore}{2} = \frac{150 + 100}{2} = 125 \text{ mm}$$

Figure 4.16 Typical bearing catalog data; series 60 single row, deep groove ball bearings.

Bearing Number	d mm	d in.	D mm	D in.	B mm	B in.	F in.	M in.	S in.	H in.	T in.	Max Fillet Radius in.	Pref. Shoulder Shaft in.	Pref. Shoulder Housing in.	Brg Wt lb	Balls No.	Balls Diam in.	Basic Static Load Rating C₀ lb	Basic Dynamic Load Rating C lb	Approx Speed Limit[2] rpm	Basic Brg No.
6000	10	.3937	26	1.0236	8	.3150	1.187	1⁷/₁₆	.120	.078	.042	.012	.472	.945	.044	7	³/₁₆	440	790	28000	6000
6001	12	.4724	28	1.1024	8	.3150						.012	.551	1.024	.049	8	³/₁₆	500	880	25000	6001
6002	15	.5906	32	1.2598	9	.3543						.012	.669	1.181	.068	9	³/₁₆	565	965	21000	6002
6003	17	.6693	35	1.3780	10	.3937						.012	.748	1.299	.089	10	³/₁₆	625	1040	19000	6003
6004	20	.7874	42	1.6535	12	.4724	2.347	2¹/₃₂	.143	.078	.065	.024	.890	1.457	.16	9	¹/₄	1000	1620	16000	6004
6005	25	.9843	47	1.8504	12	.4724						.024	1.087	1.654	.18	10	¹/₄	1110	1740	14000	6005
6006	30	1.181	55	2.1654	13	.5118	2.552	2⁹/₆₄				.039	1.346	1.929	.26	11	⁹/₁₆	1550	2290	12000	6006
6007	35	1.3780	62	2.4409	14	.5512						.039	1.543	2.205	.35	11	⁹/₁₆	1910	2760	10000	6007
6008	40	1.5748	68	2.6772	15	.5906	3.024	3¹¹/₃₂	.159	.094	.065	.039	1.748	2.441	.43	12	⁵/₁₆	2090	2900	9200	6008
6009	45	1.7717	75	2.9528	16	.6299						.039	2.008	2.717	.54	13	¹¹/₃₂	2730	3630	8400	6009
6010	50	1.9685	80	3.1496	16	.6299	3.417	3⁵/₆₄				.039	2.205	2.913	.58	14	¹³/₃₂	2940	3770	7700	6010
6011	55	2.1654	90	3.5433	18	.7087						.039	2.441	3.268	.85	13	¹³/₃₂	3820	4890	6900	6011
6012	60	2.3622	95	3.7402	18	.7087	3.811	4³/₈	.204	.109	.095	.039	2.638	3.465	.92	14	¹³/₃₂	4110	5090	6500	6012
6013	65	2.5591	100	3.9370	18	.7087						.039	2.835	3.661	.96	15	¹⁵/₃₂	4410	5280	6100	6013
6014	70	2.7559	110	4.3307	20	.7874	4.402	4²⁵/₃₂				.039	3.031	4.055	1.33	14	¹⁵/₃₂	5480	6580	5500	6014•
6015	75	2.9528	115	4.5276	20	.7874						.039	3.228	4.252	1.41	15	¹⁵/₃₂	5870	6830	5200	6015
6016	80	3.1496	125	4.9213	22	.8661			.204	.109	.095	.059	3.425	4.646	1.87	14	¹⁷/₃₂	7030	8240	4800	6016
6017	85	3.3465	130	5.1181	22	.8661	4.930	5¹/₂				.059	3.622	4.843	1.96	15	¹⁷/₃₂	7540	8560	4600	6017
6018	90	3.5433	140	5.5118	24	.9449			.218	.109	.109	.059	3.898	5.157	2.56	14	¹⁹/₃₂	8770	10000	4300	6018
6019	95	3.7402	145	5.7087	24	.9449						.059	4.094	5.354	2.67	15	¹⁹/₃₂	9410	10500	4100	6019
6020	100	3.9370	150	5.9055	24	.9449	5.718	6⁹/₃₂	.250	.141	.109	.079	4.291	5.551	2.76	15	¹⁹/₃₂	9410	10400	3900	6020
6021	105	4.1339	160	6.2992	28	1.0236						.079	4.528	5.906	3.51	15	²¹/₃₂	11500	12500	3700	6021
6022	110	4.3307	170	6.6929	28	1.1024	6.443	7³/₁₆	.261	.141	.120	.079	4.724	6.299	4.32	14	²³/₃₂	12900	14200	3500	6022•
6024	120	4.7244	180	7.0866	28	1.1024			.261	.141	.120	.079	5.118	6.693	5.34	15	²³/₃₂	13800	14700	3200	6024•
6026	130	5.1181	200	7.8740	33	1.2992						.079	5.512	7.480	8.16	15	¹³/₁₆	17600	18400	2900	6026
6028	140	5.5118	210	8.2677	33	1.2992	6.837	7¹⁹/₃₂				.079	5.906	7.874	8.62	16	¹³/₁₆	18800	19000	2700	6028•
6030	150	5.9055	225	8.8583	35	1.3780						.079	6.378	8.386	10.6	16	⁷/₈	21800	21800	2500	6030

$$C = \text{basic dynamic load rating from table}$$

$$\frac{C}{P} = \frac{\text{basic dynamic load rating from table}}{\text{equivalent load (from formula below)}}$$

$$P = XVF_r + YF_a$$

where X = a radial factor given below
V = a rotation factor = 1.0 for inner ring rotating in relation to the load
= 1.2 for outer ring

Y = a thrust factor given below
F_r = the radial load, calculated
F_a = the thrust load, calculated
e is a reference value given in the table.

When $\dfrac{F_a}{C_o}$ is smaller than or equal to e use X = 1 and Y = 0

When VF_r is greater than e use X = 0.56 and Y from table

$\dfrac{F_a}{C_o}$ =	0.014	0.028	0.056	0.084	0.11	0.17	0.28	0.42	0.56
e =	0.19	0.22	0.26	0.28	0.30	0.34	0.38	0.42	0.44
Y =	2.30	1.99	1.71	1.55	1.45	1.31	1.15	1.04	1.00

C_o = basic static load rating from table.

[2] This refers to oil lubrication and moderate load. With grease lubrication it is generally not practical to use speeds higher than ⅔ of those shown.

Figure 4.16 Typical bearing catalog data; series 60 single row, deep groove ball bearings. (Courtesy of SKF Industries, Inc., King of Prussia, Pa.)

From Figure 4.15b the minimum required viscosity v_1 is 8 cSt. The 200 SUS oil viscosity v is converted to 40 cSt using the scales on Figure 4.15b. Note that the relationship of SUS to centistokes is not linear; therefore, the v/v_1 ratio used in Figure 4.15a must be calculated in centistokes. The v/v_1 ratio is 40/8 = 5 and from Figure 4.15a the life adjustment factor a_{23} is 4 for a standard steel. The L_{10} life, therefore, is four times the calculated catalog life of 2647 hr. If a vacuum-melted steel is used, the life adjustment factor can be as high as 7. The bearing DN value is 0.25×10^6; therefore, speed effects will not significantly downgrade the life rating.

Permissible Speeds

Note that the catalog listing includes a speed limit for the 6020 bearing, 3900 rpm. This limit is based on permissible lubricant operating temperature and is set somewhat below 250°F. Speed limits listed in bearing catalogs are not absolute limits and by using proper design practices significantly higher operating speeds are possible. The areas critical in high-speed operation are lubrication, cooling, retainer design, and centrifugal effects.

Temperature Effects

When operating bearings at high temperatures, two potential problem areas are phenolic retainers and steel stability. Phenolic retainers are temperature limited and should not be used when operating temperatures approach 250°F. The standard bearing steels are stabilized at 250°F and will change dimensionally if operated at higher temperatures. It is possible to procure bearings stabilized at temperatures above 250°F; however, this will reduce hardness and fatigue life. It is also possible to procure high-hot-hardness steels such as M50, which have excellent fatigue properties at temperatures up to 500°F.

Static Load Rating

The life load relationship would indicate that at zero speed a bearing has infinite load capacity. Of course, this is not so. One could consider the limiting static load, the load at which the bearing will fracture; however, at much lower levels of loading, permanent deformations develop in the surfaces of the contacting elements. The static load rating C_0 is dependent on the magnitude of permanent deformation that can be allowed. Experience has shown that when deformations become as large as 1/10,000 of the diameter of the rolling element, objectionable noise and vibration occur during subsequent rotation. The static load rating given in Figure 4.16 is the load that corresponds to this magnitude of permanent deformation. The fracture load is approximately eight times the C_0 value.

Stationary bearings subject to vibration may exhibit pitting at very light loads. This is a form of fretting corrosion and can affect the fatigue life of bearings that alternately rotate and are stationary, such as those used on overrunning clutches.

Prorating Bearing Loads

To calculate bearing L_{10} life an equivalent load is required; however, bearings often operate under a varying load schedule. For instance, a typical operating schedule for a bearing application follows:

Load point	Time (min)	Load (lb)	rpm
1	7	300	1000
2	2	500	2000
3	1	700	3000

To find an equivalent load for the C/P term, the schedule can be prorated as follows:

$$P_E = \left[\sum \frac{N_i(P_i)^3}{N_T} \right]^{1/3}$$

where

P_E = equivalent load, lb
N_i = number of cycles at each load point
P_i = load at each point, lb
N_T = total number of cycles

For the example:

$$P_E = \left[\frac{7(1000)(300)^3 + 2(2000)(500)^3 + 1(3000)(700)^3}{7(1000) + 2(2000) + 1(3000)} \right]^{1/3}$$

$$= 497 \text{ lb}$$

Each 10-min cycle consists of 14,000 revolutions. If the C/P value is calculated to be 2, the life according to Figure 4.13 will be 8 million revolutions or 571 ten-minute cycles. The equivalent load is sometimes called the cubic mean load and the exponent used is the same as for the life formula: 3 for ball bearings and 3.3 for roller bearings. When using this method of prorating, the difference between the equivalent loads obtained with the two exponents is usually negligible; therefore, the exponent 3 is used for both ball and roller bearing calculations.

Bearing Dimensions and Tolerances

The Antifriction Bearing Manufacturers Association has arrived at standardized dimensions for rolling element bearings. These dimensions are the bore size, the outside diameter, and the width, conventionally given in millimeters. The variety of different sizes is limited so that bearings can be produced in large quantities, thereby satisfying the user's need for economical yet high-quality bearings. The basic dimension in the system is the bearing bore, which defines the shaft size upon which it is assembled. For a given bore a variety of outside diameters and widths are available in increments such that requirements for load capacity or envelope dimensions are satisfied. Through the dimensioning system, bearings from different manufacturers or of different types can be interchanged. It must be remembered, however, that although the bore, outside diameter, and width of two bearings are the same, the internal geometry may be quite different. This can affect the bearing capacity and operation. Manufacturers also offer a variety of options, such as integral seals for grease-packed bearings, snap ring grooves, and so on.

The Antifriction Bearing Manufacturers Association has set up a series of quality classes known as grades ABEC-1, 3, 5, and 7. ABEC-1 is the standard quality to which catalog bearings conform. The higher numbers correspond to better quality: that is, smaller tolerances on the bore, outside diameter, and width dimensions, as well as closer control of eccentricities, parallelism, and squareness. Also, the higher classes will have better raceway surface texture and the variation in ball or roller size within a bearing will be progressively less. Grade 3 bearings are selected from the standard grade 1 production by inspection. Grades 5 and 7 are manufactured separately with higher-precision processing. The higher-accuracy grades are used when there are requirements for especially smooth operation with low noise and vibration. An example of such an application is a machine tool spindle. Also, high-speed bearing applications should incorporate more precise bearings. Gearbox shafts operating above 3600 rpm or 5000 fpm peripheral speed should be mounted on ABEC-5 bearings.

Internal Clearance

The internal clearance of a bearing is defined as the total distance one bearing ring can move in relation to the other ring in a radial direction under no load. Ideally, bearings should operate with little or no radial clearance; however, this condition is difficult to attain. The designed bearing clearance must take into account the reduction in clearance if one or both of the raceways is mounted with a press fit. Pressing on an inner bearing ring will reduce the internal clearance 50 to 80% of the amount of the interference fit. Also, in operation the inner ring will tend to be at a higher temperature than the outer ring. The outer ring runs cooler since heat is more easily dissipated through the bearing housing to the atmosphere than from the inner ring to the shaft. This temperature differential means the inner ring expansion will reduce the bearing clearance. Catalog bearings have standard clearances which are set such that the bearing will operate properly when one of the raceways is mounted with a standard press fit. For unusual conditions bearings are available with internal clearances smaller and greater than standard. Bearing clearance will increase with bearing size and the clearances, together with recommended press fits, are tabulated in bearing catalogs.

Shaft and Housing Tolerances

Possible fits of inner races on shafts and outer races in housings range from loose to transition to tight. One of the most important factors in determining suitable fits at the bore and outside diameter is the relationship of the load to the bearing rings. The load can be either stationary or rotating with respect to the rings. For instance, in a gear shaft system using bearings with rotating inner rings the load will always act in the same direction; therefore, the inner ring rotates in relation to the load and all points on the inner race come under load each revolution. If the inner ring were loose on the shaft, the relative motion of the load would cause the ring to creep around the shaft, creating wear, fretting corrosion, and the possibility of crack initiation. In the case of a stationary load and rotating inner ring, therefore, the ring must have a tight fit on the shaft such that no clearance exists and none can be developed by the action of the load. The higher the load, the greater the press fit required. The outer ring, which in this case is stationary with respect to the load, can be slightly loose in the housing to ease assembly and disassembly. Clearance should be minimal, for the following reasons:

1. A large clearance would allow the ring to cock at assembly.
2. A large clearance would reduce the centering ability of the bearing.
3. The outer ring can deform under load if not securely assembled in the housing.

If the outer ring rotates and the load is stationary, the outer ring should be assembled with a press fit to keep it from spinning. Quite often in shaft systems there is a rotating load superimposed on stationary loads due to unbalance. This rotating load tends to make loose-fitting outer races creep. A solution to this problem is to restrain the outer ring from rotating with a pin or other locking device.

    The bearing shaft ideally should have a minimum hardness of Rc40 and be ground to size. If these requirements are not practical in a particular application, a tighter fit than normal should be used. In many applications bearing inner rings are clamped axially. The axial restraint should not be counted on to restrain the bearing from creeping and should not be used in place of a press fit.

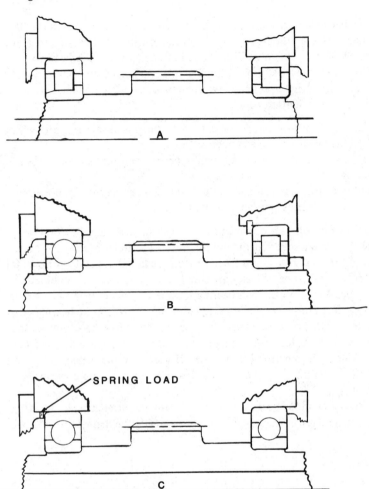

Figure 4.17 Typical bearing mounting configurations.

Bearing Mounting

The bearing arrangement must offer radial support and axial location. Also, it must accommodate thermal expansion and deflections due to load. Figure 4.17 shows three common design solutions for the support of a gear shaft. Figure 4.17A illustrates a spur gear that generates no axial loads supported by two roller bearings. The shaft is axially located by the bearing outer race shoulders and one shoulder on each inner race. It is extremely important in this type of design to allow sufficient endplay to accommodate thermal expansion of the shaft. A detailed tolerance stack-up must be made to ensure endplay under the worst dimensional conditions. A potential problem with this type of design at high speeds is bearing skidding. This condition occurs when the bearings are unloaded and do not have sufficient tractive force to keep them rolling. Skidding will result in scoring and wear and may progress to complete bearing failure. Some solutions for skidding problems are:

1. Reduction in bearing clearance such that the bearing always operates with a slight preload giving sufficient tractive force.
2. Use of out-of-round outer races which impart a constant load on the bearing in the zone where the rollers are pinched.
3. Reduction of forces that tend to retard rolling. In some cases lubrication is excessive. Retainer design can alleviate the problem.

Figure 4.17B has a ball bearing which can react thrust in either direction on one end of the shaft and a roller bearing on the other. The roller bearing has no shoulders on the inner race and is free to accommodate thermal distortions. In this case the gear might be helical. Care should be taken that the combination of thrust and radial load on the ball bearing does not result in insufficient life. Figure 4.17C illustrates a two-ball-bearing arrangement where the bearings are preloaded by a spring or wave washer. This arrangement has the advantage that at full speed and light load the spring ensures some bearing preload and the chance of skidding is minimized. For some axial distance the spring rate is sufficiently linear such that thermal expansion is accommodated and does not affect bearing loading.

For heavy combined radial and thrust loads the arrangement shown in Figure 4.18 can be used. The relief around the outside the diameter of the ball

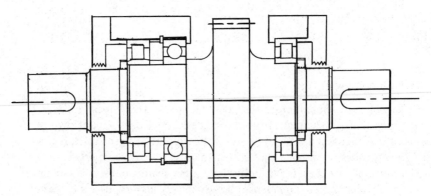

Figure 4.18 Mounting configuration for combined radial and thrust loads.

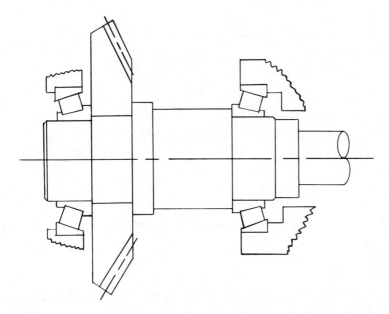

Figure 4.19 Tapered roller bearings in straddle-mounted configuration.

bearing ensures that all radial loads are reacted by the roller bearings and the ball bearing reacts only thrust. The rollers are free to move axially to compensate for temperature effects.

Tapered Roller Bearing Mounting

Figures 4.19 and 4.20 show two common ways to mount bevel gears in tapered roller bearings. Figure 4.19 illustrates a straddle mount where the gear is supported on either side. In the case shown the bearings are set up in a face-to-face configuration. In Figure 4.20 the gear is overhung mounted. A pair of tapered roller bearings takes thrust in one direction while a single bearing reacts thrust in the opposite direction. The bearings are set up in a back-to-back configuration. Care must be taken in the assembly to provide sufficient endplay for thermal expansion of the shaft. In calculating thrust loads on a shaft supported by tapered roller bearings it must be remembered that due to the angle of contact, a radial load on one bearing induces an axial load that must be reacted either by an external load or the other bearing. The induced axial load depends on the bearing internal geometry, and methods of calculation are included in catalogs of tapered roller bearings.

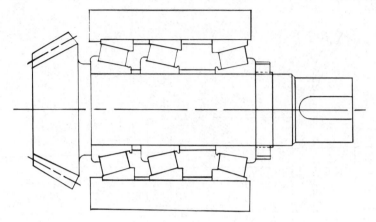

Figure 4.20 Tapered roller bearings in overhung-mounted configuration.

Bearing Lubrication

The type of lubrication to which rolling element bearings are exposed in gear units is dictated by the gear oil requirements. Gear boxes discussed in this book typically incorporate oil splash or jet lubrication systems and the bearing lubrication discussion will be limited to these areas. The types of oil used are discussed in Chapter 5.

The functions of a rolling element bearing lubricant are:

1. Provide an oil film between contacting elements.
2. Provide a cooling medium.
3. Protect bearing surfaces against corrosion.

Very little oil is required to maintain a satisfactory oil film; however, for high-speed highly loaded bearings significant oil flow must be supplied to perform the cooling function.

In a splash lubrication system the gearbox is filled with oil to a level that the gears and bearings are dipping into the lubricant as they rotate. At low speeds the heat generated can be dissipated through the casing and the unit reaches a satisfactory equilibrium temperature. Typical of such applications are electric or hydraulic motor speed reduction units. If heat generation becomes excessive the casing can be air or water cooled and an integral splash lubrication system may still be satisfactory. As speeds and loads increase a circulating system may be necessary with an external oil cooler.

In high-speed boxes, gear pitch line velocities of 5000 fpm or more, jet lubrication is used. At high speeds it is impractical to allow the components to rotate through an oil bath since the churning would create excessive heat and power loss. In some high-speed, light-load applications oil mist lubrication can be used to avoid churning losses in the bearing.

With jet lubrication one or more jet streams are directed on the bearing at the gap between the retainer and race. It is good practice to use more than one jet to ensure against oil starvation if a jet clogs. The jet diameter should be a minimum of 0.030 in. also to avoid clogging.

Flow through an oil jet can be calculated as follows:

$$Q = KA \sqrt{\frac{2g \, \Delta p}{w}}$$

where

Q = oil flow, in.^{3}/sec
A = jet area, in.^{2}
K = discharge coefficient (assume 0.65)
g = acceleration due to gravity, 386 in./sec^2
Δp = pressure drop across orifice, lb/in^2
w = specific weight, lb/in.^{3}

In order to calculate the amount of flow required to cool the bearing an estimate of the heat generated is needed. Through experience, empirical methods have been generated which approximate the losses due to friction and oil churning. Two such methods will be presented. The first method [6] estimates losses as

$$T = fRW$$

where

T = frictional torque, in.-lb
f = friction coefficient
R = shaft radius, in.
W = load, lb

The following table lists friction coefficients for various bearing configurations.

Bearing type	Friction coefficient
Radial ball bearings	0.0015
Self-aligning ball bearings	0.0010
Angular-contact bearings	0.0013
Pure thrust ball bearings	0.0013
Cylindrical roller bearings	0.0011
Spherical roller bearings	0.0018

From the friction torque the horsepower or heat generation can be calculated:

$$HP = \frac{TN}{63,025}$$

where N is the bearing rpm.

$$Q = HP(42.44)$$

where Q is the heat generated in Btu/min. This method does not consider the amount of oil flow or the oil viscosity directly. Another empirical procedure that includes these parameters is [7]:

$$Q = B[(DN)^{1.5}W^{0.07}M^{0.42}\mu^{0.25}]$$

where

Q = heat generated, Btu/min
D = bearing bore, mm
N = bearing rpm
W = bearing load, lb
μ = dynamic viscosity, Reyns (lb-sec/in.^{2})
M = oil flow, lb/hr

The following table lists values for the coefficient B:

Type of bearing	B coefficient
Angular contact	10.1×10^{-7}
Radial ball	4.46×10^{-7}
Cylindrical roller	6.46×10^{-7}

For a given flow and heat loss the oil temperature rise across the bearing can be calculated:

$$\Delta T = \frac{Q/M(c_p)}{60}$$

where

ΔT = temperature rise, °F
M = oil flow, lb/hr
c_p = specific heat, approximately 0.5 Btu/lb-°F

The conversion from GPM to lb/hr for oil is approximately

$$1 \text{ GPM} \cong 450 \text{ lb/hr}$$

A

B

C

Figure 4.21 Progression of fatigue spalling. A. Incipient fatigue spalling. B. More advanced spalling. C. Greatly advanced spalling. (Courtesy of SKF Industries, Inc., King of Prussia, Pa.)

Bearing Failures

Rolling element bearing failures are characterized by one or more of the following operating conditions:

1. Excessive vibration
2. Excessive noise
3. Overheating
4. Chip generation
5. Hard turning or excessively loose shafts

These problems can be a result of a great variety of errors but can usually be categorized as either load or lubrication related.

Lubrication-related failures occur when the supply of lubricant is insufficient or misdirected. The mating bearing surfaces, without a proper oil film, come into intimate contact, resulting in wear, heat generation, and thermal expansion. The thermal expansion reduces the bearing internal clearance and therefore increases the bearing load and the failure becomes self-perpetuating. Quite often a failure that appears to be due to overload masks a lubrication problem that led to loss of internal clearance and internally generated loading.

Load-related failures can be a result not only of the operating loads but also of forces applied during assembly. Also, the effect of operating loads can be magnified due to defective bearing seats on shafts or in housings or misalignments.

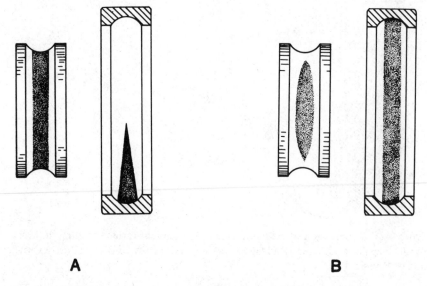

A   B

Figure 4.22 Normal radial load-bearing patterns. (Courtesy of SKF Industries, Inc., King of Prussia, Pa.)

The mode of failure related to load that is most common is fatigue spalling. Figure 4.21 shows the progression of fatigue spalling. The small spall in Figure 4.21A began with a crack which probably originated below the race surface. This initial spall will advance to the size shown in Figure 4.21B, at which point the vibration and noise level of the unit will have increased significantly. Left unattended, the spalling will progress in proportion to speed and load as shown in Figure 4.21C.

Sometimes a clue to the cause of bearing problems can be found by examination of the pattern of the load path on the bearing races. Figure 4.22 shows the normal load pattern of a radial bearing. Figure 4.22A illustrates a rotating inner ring operating with a stationary load. The pattern of a rotating outer ring with a stationary load or a rotating inner ring with a load rotating in phase is shown in Figure 4.22B. Figure 4.23 illustrates load patterns on a thrust bearing. Figure 4.23A is a normal pattern which stays within the raceway, not reaching out to the edge. In Figure 4.23B the thrust load is excessive and the balls contact the race edge, resulting in a load concentration. Figure 4.23C illustrates a normal contact pattern for a bearing experiencing combined radial and thrust load. Figure 4.24 shows load patterns that reflect problems. Figure 4.24A illustrates a bearing that is internally preloaded. This may be due to loss of clearance because of excessive press fits on the shaft or in the housing or possibly thermal expansion. Figure 4.24B shows the load pattern produced by an out-of-round housing pinching the bearing outer ring. Figure 4.24C shows the load zone when the outer ring is misaligned relative to the shaft, and Figure 4.24D the pattern when the inner ring is misaligned with respect to the housing.

Two types of pitting failures that occasionally occur and are hard to diagnose are false brinelling and electrical pitting. False brinelling is a condition caused when the gear unit is subject to vibration while the shafts are not rotating, such as in transit during shipping. It is usually characterized by polished depressions spaced equal to the distance between rolling elements. False brinelling is a form of fretting corrosion.

Electrical pitting occurs when a current seeks ground by passing through a bearing. The current is broken where the balls or rollers contact the raceway and arcing results, causing high temperatures and pitting damage.

Bearing Costs

Several variables influence the cost of gearbox bearings:

1. Obviously, when buying bearings in production quantities rather than prototype, lower prices prevail. The quantity can also determine whether the bearings are bought from a manufacturer, distributor, or supply house, which can affect the price significantly.

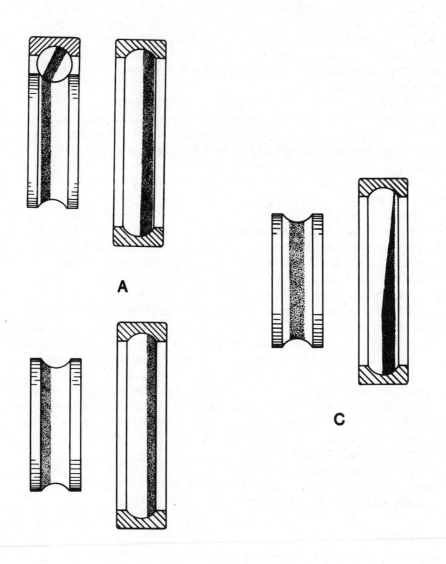

Figure 4.23 Normal thrust load-bearing patterns. (Courtesy of SKF Industries, Inc., King of Prussia, Pa.)

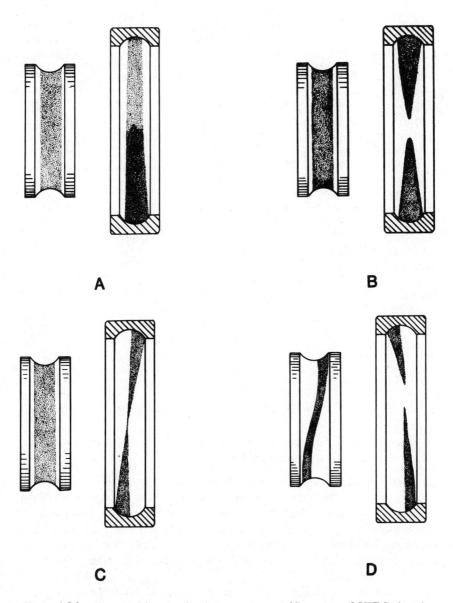

Figure 4.24 Abnormal bearing load zone patterns. (Courtesy of SKF Industries, Inc., King of Prussia, Pa.)

Table 4.1 Cost Comparison, 75-mm Bearing

Bearing type	Basic load rating (lb)	Limiting speed (rpm)	Relative cost	Cost of basic load rating (lb/dollar)
Self-aligning ball bearing	13,700	4,200	1.69	810
Single-row, deep groove ball bearing	19,600	4,200	1.32	1.480
Angular contact (40°) ball bearing	21,500	4,200	2.10	1,030
Maximum capacity, deep groove ball bearing	25,700	4,200	1.52	1,700
Cylindrical roller bearing	36,500	4,200	3.53	1,030
Double row, deep groove ball bearing	30,200	2,900	2.87	1,050
Tapered roller bearing[a]	50,100	2,460	1.00	5,010
Spherical roller bearing	68,000	2,500	4.00	1,700

[a]Bore: 3.000 in., O.D.: 5.909 in.
Source: Ref. 8.

2. Design options such as special material or processing add greatly to the basic cost.
3. Higher-than-standard precision requirements or internal clearances which are nonstandard have a significant price effect.

An interesting study was conducted investigating the comparative cost of bearings on a basic load rating per dollar basis [8]. Various configurations of the same two sizes, 55- and 75-mm bore, were evaluated to determine how much load capacity each bearing offers per dollar of cost. From Tables 4.1 and 4.2 it is seen that tapered roller bearings offer the best value on this basis. Of course, many applications have design conditions which dictate that other types of bearings are required.

It must be remembered that the initial bearing cost is not the only expenditure to be considered. If failure occurs, the expense of disassembly and reassembly must be considered, and also the downtime cost.

Table 4.2 Cost Comparison, 55-mm Bearing

Bearing type	Basic load rating (lb)	Limiting speed (rpm)	Relative cost	Cost of basic load rating (lb/dollar)
Self-aligning ball bearing	4,630	6,500	1.99	982
Single-row, deep groove ball bearing	7,500	6,500	1.42	2,220
Angular contact (40°) ball bearing	8,010	6,500	2.39	1,411
Maximum capacity, deep groove ball bearing	9,830	6,500	1.70	2,430
Cylindrical roller bearing	10,300	6,500	5.70	762
Double row, deep groove ball bearing	11,400	4,500	2.92	1,650
Tapered roller bearing[a]	15,000	3,720	1.00	6,330
Spherical roller bearing	19,300	4,000	6.02	1,350

[a]Bore: 2.1653 in.
Source: Ref. 8.

Journal Bearings

A simple radial bearing is shown in Figure 4.25. At rest, the shaft (journal) will lay on the bottom of the bearing with a load W, the shaft weight. At this point the lubricant is squeezed out and metal-to-metal contact occurs (Figure 4.26A). As the shaft begins to rotate (Figure 4.26B) it will climb up the bearing wall and be slightly off-center. The rotation of the shaft will tend to pull oil into the interface between the shaft and bearing, a wedge-shaped zone. The inlet to this converging region would like to take more oil in than the outlet will allow out. The jamming of the fluid into the converging region creates a pressure distribution between the journal and bearing as shown in Figure 4.27. The high pressure in the center forces the fluid to slow down at the inlet and speed up at the outlet, so that the flow coming in equals the flow going out. The pressure generated creates a lifting force which separates the journal and bearing with an oil film (Figure 4.26C). This pressure is the basis of hydrodynamic lubrication and the load capacity of the bearing depends on the hydrodynamic pressure

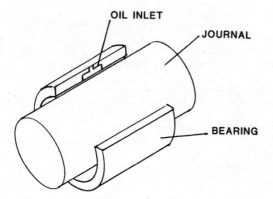

Figure 4.25   Simple radial journal bearing.

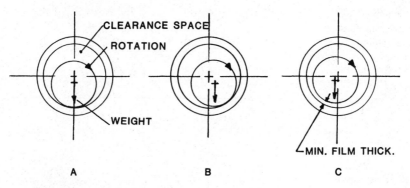

Figure 4.26   Journal bearing operation.

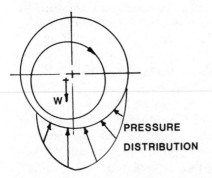

Figure 4.27   Hydrodynamic pressure distribution.

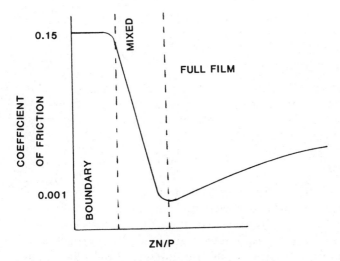

Figure 4.28 Regimes of lubrication.

generated. Increasing the fluid viscosity or the shaft velocity will increase the hydrodynamic pressure and therefore increase the film thickness. Typical hydrodynamic bearings operate with film thicknesses in the range 0.0005 to 0.002 in.

As a bearing accelerates from rest it passes through three regimes of lubrication. The first regime is boundary lubrication, where there is metal-to-metal contact. The second is mixed lubrication, where there is a transition from boundary, to hydrodynamic lubrication, where the bearing operates with a full fluid film. These regimes are illustrated in Figure 4.28, where the coefficient of friction is plotted against a bearing parameter:

$$\text{Bearing parameter} = \frac{ZN}{P}$$

where

Z = viscosity, cP
N = shaft rpm
P = unit loading, psi

The unit loading is defined as

$$P = \frac{W}{LD}$$

where

W = radial load, lb
L = bearing length, in.
D = bearing diameter, in.

The minimum ZN/P value to assure full film lubrication varies with the application. As a rule of thumb, for lightly loaded high-speed bearings with low-viscosity oils it should exceed 150. Low-speed, high-load applications with heavy oils may have ZN/P values as low as 10.

Pressure-Fed Bearings

Gearbox journal bearings are usually supplied oil under pressure. This assures a reliable source of lubricant and also provides a cooling medium. Feed pressures are typically in the range 20 to 50 psig. The oil flow to the bearing must be sufficient to replace oil lost by leakage at the bearing ends. Usually, the feed oil passages are sized sufficiently large such that the flow is metered by the bearing itself. There are various designs of pressure-fed radial bearings in use, the difference being in the shape and location of the oil feed grooves and the shape of the bearing bore.

Oil distribution grooves should not be located in the load zone since this will disrupt the hydrodynamic pressure distribution and reduce the load-carrying capacity.

A problem that arises with high speed bearings when they are lightly loaded is oil whip or whirl. It is a vibration phenomenon which occurs at either half the rotational speed or the natural frequency of the shaft when the shaft is operating well above the natural frequency. One explanation of whirl is that a wedge of oil is traveling around the bearing at the average oil velocity (one-half shaft velocity) or at the shaft critical velocity. Bearing bore designs other than the simple full round bearing (Figure 4.25) are meant to minimize this whirl tendency.

Following are descriptions of some commonly used gearbox radial bearings:

1. *Axial groove* (Figure 4.29A). This type of bearing has a cylindrical bore and one or more axial oil spreader grooves extending almost the full length of the bearing. It has high load capacity but at light loads, usually under 150 psi, is quite susceptible to oil whip.
2. *Elliptical* (Figure 4.29B). The elliptical bearing has a large clearance in one direction and a smaller clearance 90° away. It is manufactured by placing shims at the split between two halves of a cylindrical bearing and then machining the bore. The shims are removed, producing a variation in clearance around the bore, which is not truly elliptical. This type of bearing

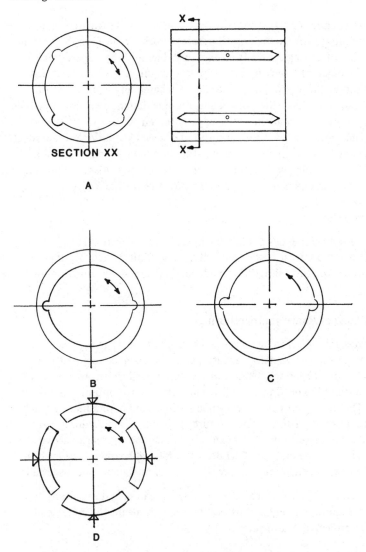

Figure 4.29 Types of pressure-fed radial bearings.

allows larger oil flows than a cylindrical axial groove design and therefore affords more cooling. It is also somewhat more stable as regards oil whip.

3. *Offset half* (Figure 4.29C). This is a modification of an axial groove bearing where the upper and lower halves are offset. It has proved successful in eliminating oil whip. A disadvantage of this design is that it can accommodate shaft rotation in only one direction.

4. *Tilting pad* (Figure 4.29D). The individual pads of this design are free to pivot and follow shaft excursions; therefore, forces produced in the bearing are incapable of driving the shaft into an unstable mode and operation is smooth. Each pad tilts such that a wedge-shaped oil film is formed which tends to center the shaft. The pivot can be located anywhere from the center of the pad to near the trailing edge. In one design the pads are pivoted near the trailing edge and the other edge is forced toward the shaft by springs. This preload adds significantly to high-speed stability. A disadvantage to this type of bearing is that it must be flooded with oil; therefore, required flows are high with accompanying power losses. Also, the bearing is mechanically complex and the parts are prone to fretting.

Relative Bearing Costs

Of the pressure-fed bearing designs the axial groove configuration is easiest to fabricate and has the lowest cost. Elliptical bore and offset half bearings are somewhat more expensive and tilting pad bearings are significantly more expensive.

Design and Rating of Radial Journal Bearings

The gearbox designer tends to be concerned with simple criteria such as the bearing unit loading, its length-to-diameter ratio, and the clearance required when defining radial journal bearings. These concerns are not sufficient to adequately design a bearing and the bearing specialists are turned to for optimization of the final design. The significant parameters to be analyzed are minimum film thickness, maximum bearing surface temperature, and maximum hydrodynamic pressure. These parameters must be traded off against one another to arrive at the optimum design. The mathematical design of journal bearings is beyond the scope of this book, but the more commonly used parameters will be discussed:

Unit Loading Typical gearbox radial bearings can operate successfully to unit loadings of approximately 550 psi. By unit load is meant the radial load divided by the projected bearing area:

$$P = \frac{W}{LD}$$

where

L = bearing length, in.
D = bearing diameter, in.

More stringent criteria are used in some cases where extreme reliability is desired. For instance, some company specifications limit unit loading to 350 psi.

Of course, the loading allowed is dependent on the bearing material. This is discussed in a later section.

Clearance Many designers as a rule of thumb use 0.001 in. of diametral clearance per inch of shaft diameter. In other words, a 6-in.-diameter journal would operate in a 6.006-in.-diameter bearing. This rule of thumb does not take shaft speed into account and a better estimate of minimum clearance is [9].

$$\text{Minimum diametral clearance} = \frac{N^{0.25}D}{6} + \frac{1.0}{D} \times 10^{-3}$$

where

N = rpm
D = shaft diameter, in.

When a bearing is pressed into a housing the bore will tend to collapse and thus affect the clearance. Clearance, therefore, should be specified after assembly or the bore made sufficiently oversize to accommodate the collapse after press fit, which may be as much as 75% of the press.

Length-to-Diameter Ratio The length-to-diameter ratio of gearbox journal bearings is usually around 1.0, although it can vary from 0.3 to 2.0. In shorter bearings there is a reduction in load-carrying capacity due to excessive end leakage. Longer bearings are susceptible to end loading due to misalignment.

Lubrication

The amount of oil leakage from both ends of a journal bearing can be estimated as follows [10]:

$$Q = 0.816 \frac{rc^3 p_s}{\mu L}$$

where

Q = total flow, gal/min
r = shaft radius, in.
μ = oil viscosity, Reyns
p_s = oil supply pressure, psi
L = uninterrupted bearing land length, in.
c = radial clearance, in.

The viscosity μ should be taken at the bearing operating temperature, which can be estimated as the average of oil temperature into and out of the gearbox. If the bearing has a full circular oil feed groove, the length L will be approximately half of the overall bearing length.

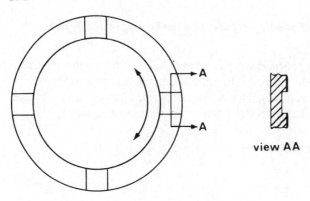

view AA

Figure 4.30   Simple thrust washer.

The oil chosen will depend on the unit speed and loading conditions. High-speed gear units typically use light turbine oils of approximately 150 SUS viscosity at 100°F. Lower-speed, highly loaded units will use heavier oils and occasionally EP additives.

After adequate oil flow, oil cleanliness is a prime factor controlling bearing life. The need for cleanliness is obvious when one considers that the magnitude of film thickness is sometimes as low as 0.0005 in. Proper cleaning of the gear unit prior to operation and continuing filtration of the lubricant cannot be too highly stressed.

Thrust Bearings

Thrust Washers   Figure 4.30 illustrates the simplest type of journal thrust bearing, a flat stationary surface which reacts thrust from a flat runner which is either integral or attached to the rotating shaft. Because there is no taper to promote hydrodynamic action, this type of bearing normally should not be designed for unit loadings above 75 psi. Higher loads are possible but require detailed design and development.

The bearing is lubricated by directing oil to the inside diameter, the oil passing through radial oil feed grooves to the outside. To simplify manufacture an even number of grooves is usually specified. Various groove designs are used, which are usually as deep as they are wide. The load-carrying lands should be approximately square.

Tapered Land   A tapered land thrust bearing is illustrated in Figure 4.31. The pad lands between oil grooves are tapered such that the runner will carry oil into a wedge-shaped region and build up a load-carrying hydrodynamic pressure. The taper may be simple or compounded, being larger at the inside

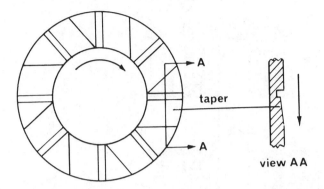

Figure 4.31 Tapered land thrust bearing.

diameter of the pad than at the outside diameter. This is done in an attempt to equalize oil flow across the pad and maintain an even temperature distribution to reduce distortion. To maintain some load-carrying ability when the bearing is starting up, the taper only extends for approximately 80% of the pad. The width-to-length ratio of the pads is approximately 1.0. Oil is fed to the inside diameter of the bearing and distributed outward through radial oil grooves. The grooves are usually dammed at the ends with a small chamfer cut to pass a

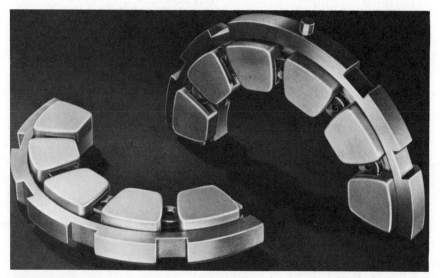

Figure 4.32 Tilting pad thrust bearing. (Courtesy of Glacier Metal Company Ltd., Wembley, England.)

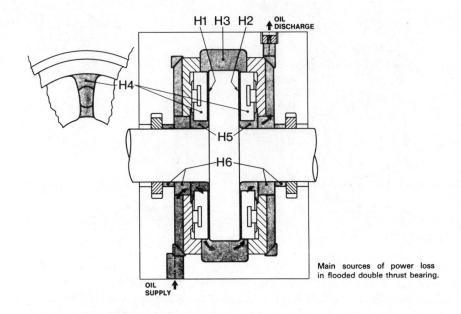

Main sources of power loss in flooded double thrust bearing.

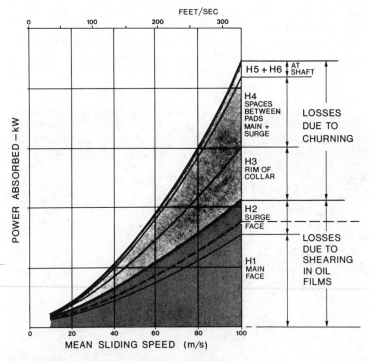

Figure 4.33 Tilting pad thrust bearing losses. (Courtesy of Glacier Metal Company Ltd., Wembley, England.)

limited amount of oil and prevent the accumulation of dirt at the groove ends. The amount of taper varies with the bearing size. A 1-in.^{2} pad may have a 0.004-in. taper, while a 7-in.^{2} pad may have a 0.008-in. taper.

Tapered land thrust bearings are used at unit loadings up to approximately 550 psi. In some critical applications where extreme reliability is required, they are derated to as low as 150 psi. Disadvantages of this type of bearing are its inability to accommodate significant misalignment and that because of the taper it has high load-carrying capacity in one direction of rotation only.

Tilting Pad   This type of bearing, sometimes called pivoted shoe or Kingsbury, consists of a number of pads which are independently free to pivot in order to provide a tapered oil film (Figure 4.32). At rest the pads are parallel to the shaft face. As the shaft rotates, an oil film is generated and each pad tilts to an angle such that the oil pressure is evenly distributed. The oil is usually fed into the center of the bearing near the shaft and exits at the bearing outside diameter. As shown in Figure 4.33, the bearing area is sealed to ensure that the bearing operates in a flooded condition. Because of this, tilted pad bearings tend to generate more heat than other types and Figure 4.33 illustrates the losses in the bearing due to shear of the oil film and also due to churning. The churning losses can be significantly reduced by directing the lubrication to the pad face only (Figure 4.34).

Tilting pad bearings can accommodate shaft rotation in both directions if the pad pivots are located in the middle of the pads. To obtain even greater tolerance for misalignment pads are sometimes mounted on leveling plates, Figure 4.35.

Tilting pad bearings can withstand unit loadings of 550 psi or more and are often specified in applications where long life and reliability are paramount. The disadvantages of this type of bearing are high cost, oil flow, and power loss.

Bearing Materials   Because a journal bearing experiences a variety of lubrication regimes from startup to full load and speed, the material must have a blend of characteristics to accommodate all the demands placed on it. This always requires a compromise between a high strength and temperature capacity and good surface characteristics, such as resistance to seizure and compliance. A discussion of the most important bearing material characteristics follows:

1. *Seizure resistance*   At startup and on occasion during operation the oil film breaks down and metal-to-metal contact between the journal and bearing will occur. The bearing material must be capable of withstanding this contact without welding, scoring, or tearing.
2. *Ability to absorb foreign objects*   If dirt particles enter the clearance space, it is desirable for the particles to become embedded in the bearing material, where they cannot score or wear the shaft.

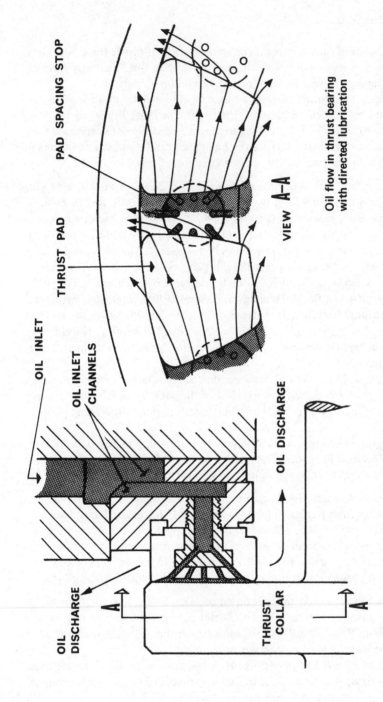

OIL INLET

OIL INLET
CHANNELS

THRUST PAD

PAD SPACING STOP

OIL DISCHARGE

OIL DISCHARGE

THRUST
COLLAR

VIEW A-A

Oil flow in thrust bearing
with directed lubrication

Figure 4.34 Directed lubrication of tilting pad thrust bearing. (Courtesy of Glacier Metal Company Ltd., Wembley, England.)

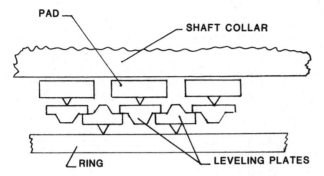

Figure 4.35 Leveling plate design for greater accommodation of misalignment.

3. *Compliability* The bearing material modulus of elasticity should be low such that if the shaft tends to touch the bearing ends due to misalignment the bearing will conform locally and the deformation will prevent severe rubbing and excessive temperature rise.
4. *Fatigue Resistance* The bearing must have the ability to withstand repeated stress cycles. In the case of gearbox bearings the load is generally steady; however, unbalance can lead to a rotating load with fatigue potential.
5. *Corrosion Resistance* If the lubricating oil contains acid or becomes acidic due to oxidation, the bearing material must be capable of withstanding the acidic attack.

The requirements of seizure resistance, embeddability, and compliability lead toward a soft bearing material such as babbitt, while fatigue strength requires a harder material. In the majority of gearbox bearings the bearing construction consists of a thin surface layer of babbitt bonded to a mild steel backing. In some cases a three-layer or trimetal construction is used with a higher-strength bearing material bonded to a steel backing and overlaid with a thin babbitt. A discussion of the most widely used materials follows:

1. *Babbitt*. Babbitts have the best seizure, embeddability, and compliability characteristics of all bearing materials. Two types are in common use; tin base and lead base. Table 4.3 lists the composition of these materials. The tin-base material is slightly more desirable in terms of seizure and corrosion resistance; however, the lead base is less expensive. A lead-base babbitt with a small percentage of tin is used in many applications as an economic compromise.

Babbitts are inherently weak and also temperature limited. They should not be used at operating temperatures over 250°F. Babbitt bearings in gearboxes are usually bonded to a mild steel backing material and the babbitt thickness is on the order of 0.030 to 0.060 in. The strength of a babbitt surface decreases with its thickness and in very high load applications such as crankshafts, which

Table 4.3 Composition of Bearing Materials

Material	Specification	Nominal composition (%)						
		Copper	Tin	Lead	Zinc	Antimony	Nickel	Arsenic
Lead-base babbitt	SAE 14	½	10	74		15		½
Tin-base babbitt	SAE 12	3½	89			7½		
Copper-lead	SAE 480	65		35				
Leaded bronze	SAE 792	80	10	10				
Leaded bronze	SAE 794	72	3	23	2			
Aluminum		1	1	6			1	

are prone to fatigue, babbitt thickness may be as low as 0.001 in.; however, in gearboxes conformability and embeddability are important and thicker layers are used.

2. *Copper-Lead.* This is the simplest of the copper-based materials and contains from 20 to 40% lead. The material has excellent fatigue strength and is capable of carrying heavy loads at high temperatures. Compared to babbitt, however, copper lead has poor seizure resistance, embeddability, and compliability. The surface behavior can be improved by using hardened journals.

3. *Bronze.* There are three categories of bearing bronze; lead bronze, tin bronze, and high-strength bronze. High-strength bronze may contain aluminum, iron, manganese, silicon, nickel, or zinc in varying percentages. Strength improves from lead to tin to high-strength bronze, and shaft compatibility characteristics deteriorate. Bronzes are cost effective since they can be easily cast and machined and do not require a steel backing.

4. *Aluminum.* Aluminum has good fatigue resistance and corrosion resistance but does not have the surface properties of babbitt. Hardened shafts should be used in conjunction with aluminum bearings.

5. *Trimetal composite bearings.* In order to take advantage of the surface characteristics of babbitt yet improve bearing strength and temperature resistance, trimetal bearings have come into use. An intermediate layer of approximately 0.020 in. of copper-lead or leaded bronze is sandwiched between a mild steel backing and an overlay of babbitt approximately 0.002 in. thick.

Shaft Definition

Many journal and thrust bearing problems are a result of shaft discrepancies such as improper geometry or poor surface finish. The important geometric characteristics are taper, out of roundness, and grinding discrepancies such as waviness, chatter, or lobing, which must be closely controlled.

Shaft taper in the bearing area should be limited to 0.0002 in. per inch of length. Out of roundness should be limited to 0.0005 in. for shaft diameters up to 5 in. and 0.001 in. for shafts above 5 in. in diameter. Shaft surface finish should be a maximum of 20 rms with a preferred finish of approximately 10 rms.

Journals operating with babbitted surfaces can be made of soft steel with a minimum hardness of 200 bhn. Aluminum and copper-lead materials can be operated with shafts of 300 bhn minimum hardness. Where loads are high a journal hardness of Rc50 is recommended. In general, it is desirable to harden shaft surfaces in any case, in order to obtain longer life and improved wear and abrasion characteristics. In gearboxes the bearing journals quite often are integral with a pinion and therefore made of the gear material and hardened to the gear tooth specification and ground. In the case where gears are assembled on a shaft, the shaft might be SAE 4140 or 4340 steel and the journals nitrided and ground.

A

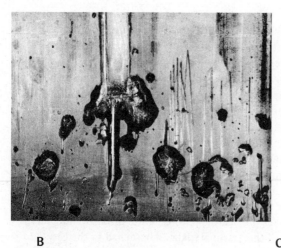

B   C

Figure 4.36 Effect of lubricant contamination. (Courtesy of Glacier Metal
Company Ltd., Wembley, England.)

Journal Bearing Failures

Unlike rolling element bearings, fatigue failures are not common in gearbox journal bearings. Failures that occur are more likely to be associated with contamination of the lubricant, insufficient lubrication, dynamic excursion of the shaft, or faulty assembly.

Lubricant contamination. Abrasive materials in the lubricant may be a result of insufficient cleaning of the machined components at assembly, dirt entering the unit through breathers or bypassed oil filters, or wear particles generated inside the unit. Figure 4.36A shows a babbitted surface which has been scored and pitted by dirt, and Figure 4.36B shows surface distress to a greater extent caused by dirt particles. Figure 4.36C illustrates a thrust pad scored circumferentially by dirt particles.

The bearing in Figure 4.36A can be cleaned and reused provided that the wear experienced is not excessive and the bearing clearance remains within specification. Of course, the source of contamination must be eliminated and the lubrication system flushed out before continuing operation.

Lubricant cleanliness cannot be too greatly emphasized when operating journal bearings. From the assembly area, where all oil passages should be mechanically cleaned to eliminate machining chips, to the test stand, where the unit should be extensively flushed, and in the field, where oil must be carefully filtered, continuous care must be taken not to contaminate the oil. Filtration should be a maximum of 40 μm, and lower filtration levels to 10 μm are beneficial. When servicing the unit in the field care must be taken not to introduce dirt through filler ports or inspection covers.

Wiping of Bearing Surfaces. When the rotating journal and the bearing metal touch during operation the rubbing causes melting and smearing, as shown in Figure 4.37. Wiping may be due to insufficient clearance, overheating which closes down the clearance, high transient loads, shaft vibration due to unbalance, or dynamic journal instabilities. Figure 4.37A shows both halves of a wiped babbitted bearing and Figure 4.37B illustrates an overlay-plated copper-lead bearing wiped on half the circumference. When wiping occurs if the surface distress is light, the bearing can be cleaned and reused provided that the wear does not result in excessive clearance. The cause of the wipe should be identified and corrected. For instance, if a vibration survey shows synchronous vibration, the shaft should be balanced. If oil whip is detected, a different profile bore is indicated.

Corrosion. Corrosion of a bearing is a result of chemical attack by reactive materials in the oil stream. The most common problem is oxidation products formed in the oil, which corrode materials such as lead, copper, cadmium, and zinc. Figure 4.38 shows the severely corroded surface of a

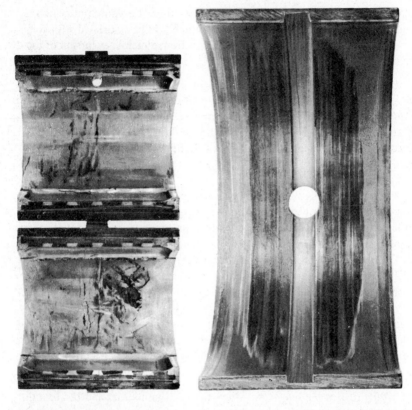

Figure 4.37 Wiping of bearing surface. (Courtesy of Glacier Metal Company Ltd., Wembley, England.)

copper-lead bearing. Lead reacts rapidly with oil oxidation products. To solve this problem oxidation inhibitors are incorporated in the lubricant and protective alloying elements such as tin are included in the lead babbitt formulation.

Faulty Assembly. The fit of the bearing in the housing is a potential source of difficulty. If the bearing is too loose in the housing fretting corrosion can result. This also can occur if the bearing housing design is too flimsy. Too tight a press fit can cause bore distortion. Another cause of bore distortion is entrapment of foreign particles between the bearing and its housing during

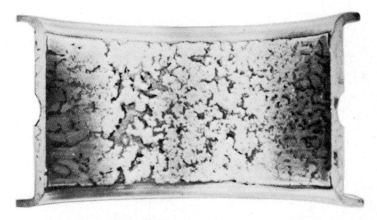

Figure 4.38 Bearing corrosion. (Courtesy of Glacier Metal Company Ltd., Wembley, England.)

assembly. This will result in a localized bore distortion and potential overheating. Misalignment of the shaft with respect to the housing will lead to uneven bearing wear, as shown in Figure 4.39.

Salvage Procedure

Following a journal bearing seizure the shaft may be scored or worn to such an extent that rework is required. There are two potential rework procedures. One is to grind the shaft down and press fit on a sleeve, which is ground to size after assembly. The sleeve must be sufficiently thick to withstand the press fit stresses and the fit should be approximately 0.001 in. per inch of shaft diameter. Another rework method is to grind the shaft down at the distressed area and build it up with chrome plate or a metal spray. Following is a rework procedure for hard chrome plating:

1. Grind the diamter 0.015 in. under the low limit.
2. Bake after grinding for 4 hr at 275°F.
3. Shot peen the rework area using 170 shot to an intensity of 0.012 to 0.014A.
4. Hard chrome plate the rework area and bake after plating for 4 hr at 275°F. Plate to 0.004 in. above the high limit.
5. Grind to finished size.

Care must be taken in grinding the chrome plate since abusive grinding can overheat the base metal, resulting in temper and cracks. These cracks are masked by the chrome plate and cannot be detected by magnaflux inspection. The cracks can propogate through the chrome plate during operation, lifting off the plating and causing bearing failure.

Figure 4.39 Bearing pattern due to misalignment. (Courtesy of Glacier Metal Company Ltd., Wembley, England.)

SEALS

Gearbox shaft seals are usually either elastomeric lip seals or noncontacting labyrinth seals. Elastomeric lip seals are generally found in low-speed applications in conjunction with rolling element bearings. The limiting surface speed of a lip seal is on the order of 3600 fpm. Somewhat higher speeds are possible but require careful design and development. When journal bearings support a shaft, the shaft movement may be excessive for successful lip seal operation.

Labyrinth seals because they are noncontacting are not speed limited and also do not introduce any friction torque into the system. They are used in critical high-speed applications because there is little likelihood of failure. On occasion, carbon face seals are incorporated in gear applications where speeds are high and a more positive seal than a labyrinth is desired.

Labyrinth and lip seals will allow oil leakage if the gearbox internal pressure is higher than the external ambient pressure. This situation may occur for the following reasons:

1. The gearbox may be vented to an area where the pressure is higher than ambient pressure. This causes the unit to be back-pressured, resulting in oil leakage.

2. On occasion the gearbox may serve as the sump for the scavenge oil of another piece of equipment, such as a turbine. This oil may be aerated and at a pressure higher than ambient.
3. The area external to the gearbox may be at a slight vacuum and the gear unit, if vented to atmosphere, will be at a higher internal pressure.
4. If the gear unit is not vented, heat generated in the box can expand the air, creating a pressure differential with the outside.

If any of the conditions noted above result in oil leakage, several solutions are possible:

1. Lip seals with greater interference or higher spring force can be used; however, this may result in excessive wear.
2. Labyrinth seals can be internally pressurized or internally drained to prevent oil leakage.
3. A face seal can be incorporated which can seal against small pressure differentials.

The following paragraphs will describe the seal configurations most commonly used in gearboxes.

Labyrinth Seals

Labyrinth seals are devices that limit leakage between a rotating shaft and a stationary housing by maintaining a close radial clearance between the two. Figure 4.40 illustrates a typical labyrinth seal where several stages of knives deter oil leakage from the gearbox cavity. The labyrinth pictured has two sets of knives with an interstage area which can be used for one of two purposes. A drain can be located between the knives such that any oil leakage past the first stages will be diverted back into the gearbox.

The interstage area can also be used to introduce air pressurization and maintain the air side of the seal at a higher pressure than the oil side. Usually, a pressure differential of 1 or 2 psi is sufficient to effectively eliminate oil leakage. This technique is used when a source of bleed air such as a turbine is readily available. An estimate of the airflow through a labyrinth seal to create a desired pressure differential can be made with the following equation [11]:

$$M = K_1 K_2 K_3 K_4 P_0 A$$

where

M = airflow, lb/sec
K_2 = coefficient dependent on the ratio of radial gap G to knife tip width W (Figure 4.41)

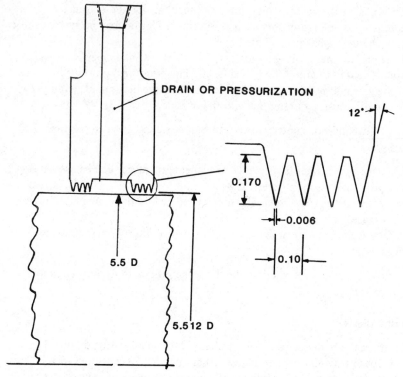

Figure 4.40 Labyrinth seal.

K_3 = coefficient dependent on the ratio of radial gap G to knife pitch P and number of knives (Figure 4.41)

K_1 = coefficient dependent on pressurization air temperature (Figure 4.41)

K_4 = coefficient dependent on the pressure ratio (low pressure/high pressure) across the seal and the number of knives (Figure 4.41)

P_0 = high pressure, psia

A = gap area, in.^{2}

As an illustration of this analysis, assume that a pressure of 2 psig (16.7 psia) is desired in the interstage area of the seal shown in Figure 4.40. The air is at a temperature of 100°F.

$$\text{Pressure ratio} = \frac{14.7}{16.7} = 0.88$$

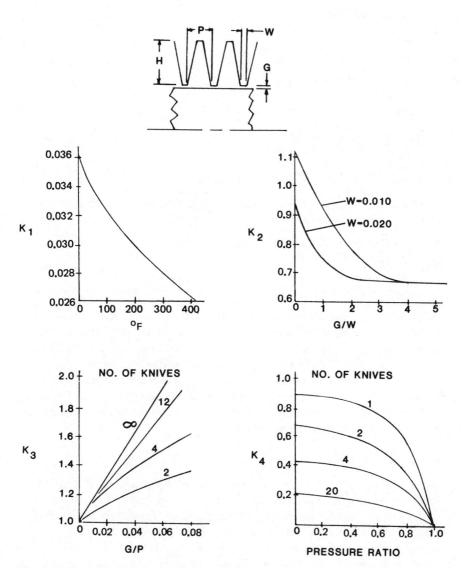

Figure 4.41   Labyrinth seal air leakage. (From Ref. 11.)

$$P = 0.10$$

$$W = 0.006$$

Number of knives $= 4$

$$K_1 = 0.0325$$

$$K_2 = 1.0$$

$$K_3 = 1.5$$

$$K_4 = 0.2$$

$$A = 0.104$$

$$P_0 = 16.7$$

$$M = 0.017 \text{ lb/sec airflow}$$

If volume flow is desired, the conversion from pounds per second to standard cubic inches per minute (SCIM) is

$$5000 \text{ SCIM} \cong 0.004 \text{ lb/sec}$$

Therefore,

$$M = 21250 \text{ SCIM} = 12.3 \text{ SCFM} \quad \text{(airflow through one-half the seal)}$$

In the interest of limiting leakage it would be advisable to reduce the clearance to the smallest possible amount. The practical clearance limit must take into account bearing tolerances, dynamic deflections, and thermal distortions. If the shaft motion exceeds the clearance, a rub will occur which will wear or score the knives and the shaft. In journal bearing gear boxes the radial clearance is in the order of 0.010 in. To achieve closer clearances it is possible to coat the knives with an abradable material such as tin babbitt. If the shaft touches, no harm will result and minimum gaps can be maintained.

Elastomeric Lip Seals

Figure 4.42 illustrates a typical elastomeric lip seal. By elastomeric it is meant that the sealing lip is made of a synthetic rubber compound. Oil sealing is accomplished through an interference fit between the flexible sealing lip and the shaft. Because the elastomer can lose tension and elasticity during operation, a garter spring is usually incorporated on the lip to ensure pressure at the sealing interface. In the design of Figure 4.42 a dirt excluder lip on the air side is used to shield the oil sealing lip from contaminants. Because the dirt exclusion lip is unlubricated its design interference will be less than the oil seal lip interference.

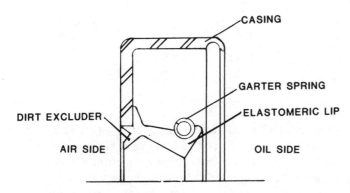

Figure 4.42 Elastomeric lip seal.

The amount of interference and lip contact pressure required depends on the following variables:

Elastomer compression set
Thermal expansion
Shaft eccentricity
Shaft-to-bore misalignment

Typical lip-to-shaft interferences, or pinches as they are sometimes called, are on the order of 0.035 in.

A paradox of lip seals is the fact that although they are sealing devices, in order to operate properly there must be a lubricant film between the lip and the shaft. Successful sealing and long seal life depend on maintenance of this film, which is usually about 0.0001 in. thick. If no film is present, the seal lip will wear out or tear. The majority of lip seal applications, about 80% will leak slightly; on the order of 0.002 g/hr or about 1 drop every 11 hr [12]. About 15% will leak 0.002 to 0.1 g/hr. This level of leakage may be unacceptable.

Shaft Specification

Medium carbon steel such as SAE 1035 or 1045 or stainless steel shafts are the best sealing surfaces, but chrome or nickel plated shafts are also acceptable. The shaft hardness should be Rc 30 minimum. Shaft finish should be 10 to 20 μin. and the shaft should be plunge-ground with no machine lead. Wear sleeves are commercially available which can be pressed over the shaft and used as a sealing surface. One advantage of wear sleeves is that they can be easily replaced in case of surface distress. Seal manufacturers' catalogs list standard shaft diameters for which lip seals are readily available. The diameter tolerances are as follows:

Shaft diameter (in.)	Tolerance
Up to 4.0	±0.003
4.001–6.0	±0.004
6.001–10.0	±0.005

The maximum eccentricity a lip seal can accommodate varies with the shaft rpm. At 1000 rpm the eccentricity can be up to 0.015 in. while at 4000 rpm eccentricity is limited to 0.007 in. Operating eccentricity is a combination of the amount by which the shaft is off center with respect to the bore and the dynamic runout. It can be measured by the total movement of an indicator mounted on the casing and held against the side of the shaft while the shaft is slowly rotated.

Bore Specification

The seal outside diameter-bore interface is a potential oil leakage path. The press fit of the seal in the housing should be a minimum of 0.004 in. with higher press fits for bore diameters above 4 in. These recommendations apply to ferrous housings. If another material such as aluminum is used, the higher coefficient of thermal expansion must be considered. The bore finish should be a maximum of 125 μin. It is possible to procure lip seals with elastomeric coatings on the outside diameter which will seal at the bore interface, but these seals must be assembled with higher press fits than seals with metal casings. It is good practice to brush a synthetic rubber coating on the bore inside diameter prior to assembly to fill minor imperfections on the bore and enhance sealing.

Lip Seal Materials

The most commonly used lip seal materials in gearbox applications are the nitriles, sometimes known as Buna N. The standard nitrile compound is compatible with most mineral oils and can operate continuously at temperatures from –65 to 225°F. Nitrile seals are relatively inexpensive and readily available. This material is not suited for temperatures above 250°F because it tends to harden and it is not recommended for highly compounded lubricants (EP additives).

For high-speed and high-temperature applications fluoroelastomers are recommended. These materials, sometimes known by the trade name Viton, can be used in a temperature range of –40 to 400°F. They have outstanding resistance to a wide variety of fluids, including synthetic lubricants such as Mil-L-7808 and Mil-L-23699. For critical applications the fluoroelastomers are undoubtedly the best technical choice. The disadvantages of this material are higher cost and limited availability of particular configurations.

Polyacrylate elastomers are used for operation with EP-type fluids. They have good resistance to temperatures up to 300°F but have poor low-temperature properties.

Silicones are applicable to a wide temperature range of –100 to 300°F and also have low friction and wear properties. They have poor resistance to oxidized oil and some EP additives.

Carbon Face Seals

Figure 4.43 illustrates a carbon face seal installation. The seal assembly consists essentially of two flat-surfaced rings, the runner rotating with the shaft and the stationary carbon member carried in the housing. The carbon sealing face must be free to follow the motions of the runner despite axial shaft movement or deviations from pure rotational movement. In order to accommodate shaft axial and radial movement, yet maintain a sealing force, the carbon sealing face is spring loaded against the runner. The spring force is basically constant over some range of axial movement and the seal must be axially positioned at assembly to be within its operating range. This may be accomplished by shimming at assembly. Because the carbon sealing face is free floating a secondary static seal is required to close the leakage path and this is usually an elastomeric O-ring. To achieve long seal life the spring force must be minimized yet adequate to overcome axial friction, inertia, and dynamic forces. It can be seen from Figure 4.43 that at the radial sealing interface fluid tends to be pumped back into the bearing cavity due to centrifugal force; thus the face seal can seal against small pressure differentials.

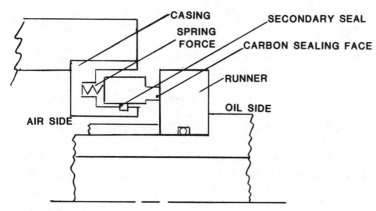

Figure 4.43 Carbon face seal.

The rubbing surfaces must be compatible such that during mechanical contact there is no galling, abrasion, or wear. Carbon is used because of its rubbing friction characteristics, compatibility with many materials, temperature resistance, low mass, low porosity, and ease of fabrication. Runners can be hardened stainless steel or plated steels such as 4140 coated with chrome or carbide. Runner hardness should be Rc 58 minimum and the runners should be ground and lapped to a flatness of three helium light bands.

Normally in a face seal installation the runner is the rotating member; however, there are some designs with stationary runners and rotating seal assemblies. This type of design is speed limited because of the effect of centrifugal force on the internal seal components.

REFERENCES

1. SKF Engineering Data, SKF Industries Inc., King of Prussia, Pa., 1973.
2. The Rolling Elements Committee of the Lubrication Division of the ASME, Life Adjustment Factors for Ball and Roller Bearings, 1971 (Library of Congress Card Number 70-179492).
3. Grubin, A., and Vinogradova, I., Investigation of the Contact of Machine Components, Moscow, Ts-NIITMASH, Book 30, 1949 (DSIR, London, Translation 337).
4. Archard, J. and Cowking, E., Elastohydrodynamic Lubrication of Point Contacts, *Proceedings of the Institution of Mechanical Engineers*, Vol. 180, Part 3b, 1965-1966, pp. 47-56.
5. SKF Catalog 310-110, Spherical Roller Bearings, SKF Industries, Inc., King of Prussia, Pa., June 1981.
6. Palmgren, A., Ball and Roller Bearing Engineering, SKF Industries Inc., King of Prussia, Pa., 1945.
7. NACA Technical Note 3003, September 1953.
8. Benes, Capacity/Cost Factor Can Help You Find Bargains in Bearings, *Machine Design*, November 2, 1972.
9. Conway-Jones, J. M., Application and Design of Plane Bearings in Power Transmission Machinery, First International Power Transmission Conference, June 1969.
10. Yahraus, W. A., Influence of Lubrication System Variables on Sleeve Bearing Performance, SAE Paper SP-390, May 1975.
11. Egli, A. The Leakage of Steam Through Labyrinth Seals, ASME Paper FSP-57-5, 1935.
12. Horve, L. A., Fluid Film Sealing, Elastomeric Lip Seals, ASLE Education Course, Chicago Rawhide Manufacturing Co., Chicago, 1961.

5
LUBRICATION SYSTEMS

The purpose of a gearbox lubrication system is to provide an oil film at the contacting surfaces of working components to reduce friction. Equally as important, the lubrication system absorbs heat generated in the gearbox and dissipates it so that component temperatures do not become excessive. Prior to discussing specific details of gearbox lubrication, it is appropriate to review lubrication fundamentals.

VISCOSITY

Friction between rubbing surfaces is reduced by separating the surfaces with a film of oil. The lower fluid friction is substituted for the frictional resistance of dry metal surfaces. Viscosity is a measure of fluid friction. Consider Figure 5.1, where plate 1 is moving with a velocity U over a stationary plate 2 and the plates are separated by an oil film of thickness h. The force F on plate 1 causes a shearing stress in the fluid. The fluid in contact with the moving plate is at velocity U and the fluid in contact with the stationary plate is at rest. Newton's law of viscous flow states that the shearing stress in the fluid is proportional to the rate of change of the velocity, and if we assume that the rate of shear is constant,

$$S_s = \frac{\mu U}{h}$$

If U is in inches per second, h is in inches, and S_s is in pounds per square inch, μ is the absolute viscosity in pound seconds per inch squared. A pound second

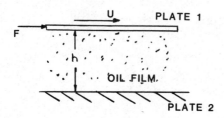

Figure 5.1 Fluid friction.

per inch squared is called a Reyn in honor of Osborne Reynolds. In the metric system the absolute viscosity is usually expressed in centipoise:

$$Z = \mu(6.9 \times 10^6)$$

where

Z = centipoise or $1/100$ dyne-sec/cm^2
μ = Reyns or lb-sec/in.^{2}

Quite often, viscosity is expressed in Saybolt seconds universal (SSU), where the viscosity is determined with an instrument that measures the time in seconds for 60 cm^3 of a fluid to pass through a standard capillary tube at a given temperature. This is a kinematic viscosity and is related to absolute viscosity as follows:

$$Z_k = \left(0.22T - \frac{180}{T}\right)$$

where

Z_k = kinematic viscosity, cSt
T = number of Saybolt seconds

The absolute viscosity in centipoise is

$$Z = \rho Z_k = \rho\left(0.22T - \frac{180}{T}\right)$$

where ρ is the specific gravity at the given temperature. Specific gravity is measured by the petroleum industry at 60°F with a glass hydrometer calibrated in degrees as specified by the American Petroleum Institute. This is an arbitrary scale which is converted to actual specific gravity as follows:

$$\rho_{60} = \frac{141.5}{131.5 + \text{degrees API at 60°F}}$$

Table 5.1 Viscosity and Specific Gravity of SAE Oils

SAE number	Saybolt seconds universal		Degrees API at 60°F	ρ at 60°F	Specific gravity, ρ	
	100°F	210°F			100°F	210°F
10	183	46	30.2	0.875	0.861	0.822
20	348	57	29.4	0.879	0.865	0.827
30	489	65	28.7	0.883	0.869	0.830
40	680	75	28.3	0.885	0.871	0.832
50	986	90	26.6	0.895	0.881	0.842
60	1394	110	26.3	0.897	0.883	0.844
70	1846	130	25.6	0.901	0.887	0.848

where ρ_{60} is the specific gravity at 60°F. The specific gravity at any other temperature is related to ρ_{60} as follows:

$$\rho = \rho_{60} - 0.00035(°F - 60)$$

Table 5.1 gives SSU and ρ_{60} values for various SAE oils. It can be seen from this table that the viscosity of oil varies widely with the operating temperature. The viscosity-temperature relationship of lubricating oils can be plotted as a straight line if the scales are correctly arranged. To determine the viscosity of an oil at

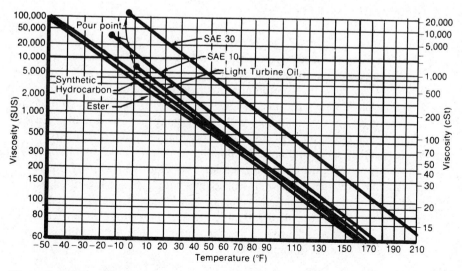

Figure 5.2 Oil viscosity versus temperature. (From Ref. 1.)

any temperature the Saybolt seconds are found at two temperatures, usually 100 and 210°F and a straight line is drawn between these points. Figure 5.2 presents the viscosities of some common fluids as a function of operating temperature.

VISCOSITY INDEX

The viscosity index is a measure of how much the oil viscosity varies with temperature. It would be ideal if the fluid viscosity were the same at low and high temperatures, but this is unattainable. Fluids that have low viscosity variations with temperature have a high viscosity index, whereas a low viscosity index defines a fluid with a widely fluctuating viscosity with respect to temperature. Typical viscosity indexes for petroleum oils range from 90 to 110.

POUR POINT

The pour point of an oil is the lowest temperature at which it will flow. This parameter must be considered when designing a unit for outdoor use; however, in most low-temperature applications the oil will be heated prior to startup. When no provision for oil heating is made, the pour point should be 20°F above the lowest expected ambient temperature.

GEAR LUBRICANTS

The selection of a gear oil depends on several factors, including the unit's operating speed and load, ambient temperatures, and which lubricants will be available at the operating site. The most important parameter in selecting a lubricant is the viscosity. High-speed units require less viscous oil than gears operating at low speed. At high speed an acceptable oil film is generated at the tooth contact even with a low-viscosity oil. Also, the churning that occurs at high speed will be less severe with a low-viscosity oil, resulting in lower power losses. At lower operating speeds a thinner oil film is generated and more viscous oils are required to separate the contacting surfaces. Also, low-speed gears are loaded to higher levels. Often, a gear unit will contain both high- and low-speed gear meshes. In this case a compromise must be struck and some development may be necessary. AGMA Standard 250.04 [2] defines a series of oils by viscosity as shown in Table 5.2. Each grade is given an American Gear Manufacturers Association (AGMA) lubricant number.

Table 5.2 Viscosity Ranges for AGMA Lubricants

Rust and oxidation inhibited gear oils			Extreme pressure gear lubricants[c]	Viscosities of former AGMA system[d]
AGMA lubricant number	Viscosity range[a] mm²/s (cSt) at 40°F	Equivalent ISO grade[b]	AGMA lubricant number	SSU at 100°F
1	41.4–50.6	46		193–235
2	61.2–74.8	68	2 EP	284–347
3	90–110	100	3 EP	417–510
4	135–165	150	4 EP	626–765
5	198–242	220	5 EP	918–1122
6	288–352	320	6 EP	1335–1632
7 Comp[e]	414–506	460	7 EP	1919–2346
8 Comp[e]	612–748	680	8 EP	2837–3467
8A Comp[e]	900–1100	1000	8A EP	4171–5098

Note: Viscosity ranges for AGMA lubricant numbers will henceforth be identical to those of ASTM 2422.
[a] Viscosity System for Industrial Fluid Lubricants, ASTM 2422. Also British Standards Institute, B.S. 4231.
[b] Industrial Liquid Lubricants—ISO Viscosity Classification. International Standard, ISO 3448.
[c] Extreme pressure lubricants should be used *only* when recommended by the gear drive manufacturer.
[d] AGMA 250.03, May 1972 and AGMA 251.02, November 1974.
[e] Oils marked Comp are compounded with 3 to 10% fatty or synthetic fatty oils.
Source: Ref. 2.

Table 5.3 Specification for R&O Gear Oils (Including Compounded Gear Lubricants)

Property	Test procedure	Criteria for acceptance		
Viscosity	ASTM D88	Must be as specified		
Viscosity index	ASTM D2270	90 min		
Oxidation stability	ASTM D943	Time to reach a neutralization number of 2.0:[a]		
		AGMA grade	Hours (minimum)	
		1, 2	1500	
		3, 4	750	
		5, 6	500	
Rust protection	ASTM D665	No rust after 24 hr with synthetic seawater		
Corrosion protection	ASTM D130	No. 1 strip after 3 hr at 120°C (250°F)		
Foam suppression	ASTM D892	Must be within these limits:		
			Max volume of foam (ml) after:	
		Temperature	5-min blow	10-min rest
		Sequence I 24°C (75°F)	75	10
		Sequence II 93.5°C (200°F)	75	10
		Sequence III 24°C (75°F)	75	10
Demulsibility	ASTM D2711	Must be within these limits:[a]		
		Max. percent water in the oil after 5-hr test	0.5%	
		Max. cuff after centrifuging	2.0 ml	
		Min. total free water collected during entire test	30.0 ml	
Cleanliness	None	Must be free from grit and abrasives		

[a]The criteria for acceptance indicated for oxidation stability and demulsibility are not applicable to compounded gear oils.
Source: Ref. 2.

218

RUST- AND OXIDATION-INHIBITED OILS

AGMA lubricant numbers 1 to 8 are petroleum-based rust- and oxidation-inhibited oils which meet certain American Society of Testing and Materials (ASTM) specifications, as shown in Table 5.3:

Oxidation Stability (ASTM D943) In this laboratory test, pure oxygen is bubbled through a mixture of oil and water in the presence of copper and iron wire catalysts at 95°C. The test life is reported as the time in hours it takes the oil to reach an acidity number of 2.0.

Rust Protection (ASTM D665) This procedure measures the rust-preventing characteristics of an oil in the presence of synthetic seawater. A steel specimen is used.

Corrosion Protection (ASTM D130) This test evaluates the ability of an oil to control the corrosion of copper and copper alloys in the presence of water at elevated temperatures.

Foam Suppression (ASTM D892) When air contaminates an oil circulation system, the efficiency of the system depends on the oil's natural resistance to foaming and its ability to break foam quickly. In this test foam is created in the test oil by blowing air through it for 5 min. The volume of foam is recorded at the end of a 10-min settling period.

Demulsibility (ASTM D2711) The degree of oil demulsibility is determined by the amount of time it takes equal portions of distilled water and oil to separate after the two have been mixed by a 1500-rpm steel paddle at a temperature of 180°F.

The lighter rust- and oxidation-inhibited oils, AGMA lubricant numbers 1 to 4, are sometimes referred to as turbine oils and are widely used in gear units, particularly those operating at high speeds.

EXTREME PRESSURE GEAR LUBRICANTS

Extreme pressure lubricants are petroleum-based oils containing special chemical additives which can increase the load-carrying capacity of gears by forming a film on the metal surfaces which provides separation when the lubrication film becomes thin enough for the asperities to contact. Some boundary films will melt at lower temperatures than others and will then fail to provide protection at the surfaces. For this reason, many extreme pressure lubricants contain more than one chemical for protection over a wide temperature range. Some of the EP additives commonly used in gear oils are those containing one or more compounds of chlorine, phosphorus, sulfur, or lead soaps. EP additives are

Table 5.4 Specification for Extreme Pressure Gear Lubricants

Property	Test procedure	Criteria for acceptance
Viscosity	ASTM D88	Must be as specified
Viscosity index	ASTM D2270	90 min
Oxidation stability	ASTM D2893	Increase in kinematic viscosity of oil sample at 95°C (210°F) should not exceed 10%
Rust protection	ASTM D665	No rust after 24 hr with distilled water
Corrosion protection	ASTM D130	No. 1 strip after 3 hr at 100°C (212°F)
Foam suppression	ASTM D892	Must be within these limits:

		Max volume of foam (ml) after:
Temperature	5-min blow	10-min rest
Sequence I 24°C (75°F)	75	10
Sequence II 93.5°C (200°F)	75	10
Sequence III 24°C (75°F)	75	10

Property	Test procedure	Criteria for acceptance		
Demulsibility	ASTM D2711 (Modified for 90 ml water)	Must be within these limits:		

	AGMA grades	
	2EP to 6 EP	7EP, 8EP
Max. percent water in the oil after 5-hr test	1.0%	1.0%
Max. cuff after centrifuging	2.0 ml	4.0 ml
Min. total free water collected during entire test (start with 90 ml of water)	60.0 ml	50.0 ml

Property	Test procedure	Criteria for acceptance
Cleanliness	None	Must be free from grit and abrasives
EP property	ASTM D2782 (Timken Test) DIN 51-354 (FZG Test)	An oil must pass both a 60-lb Timken OK load, and 11 stages on the FZG machine with A/8.3/90°C parameters for acceptance.
Additive solubility	None	Must be filterable to .25 μm (wet or dry) without loss of EP additive

Source: Ref. 2.

chemically reactive and care must be taken when they are used that metals such as zinc or copper which may be in the gear unit are not attacked. An existing unit should not be changed from a straight mineral oil to an EP oil without the manufacturer's approval.

Lubricant numbers 2EP to 8EP in AGMA Standard 250.04 [2] are EP oils which meet certain ASTM specifications, as shown on Table 5.4. Most of the tests are as those described for Table 5.3, however, for EP properties a Timken test and a FZG test are specified.

In the Timken test a rectangular steel block is brought into contact with a rotating steel cylinder to determine the maximum load a lubricant will carry before its film strength is exceeded. The load is increased until the block shows surface distress, which indicates lubricant failure. The FZG test uses operating gears to measure wear and surface distress. This is a back-to-back test with two connected gears operating at 1760 rpm and two connected 2640-rpm pinions. Load is applied by loosening the coupling in the pinion shaft and placing weights on a load arm. The preload is locked into the system by tightening the bolts of the coupling, and the load arm and weights are removed before starting the machine. Twelve uniform weights are consecutively applied and after each run the gears are checked for weight loss and visual condition. When a 10-mg weight loss is recorded between two runs the oil is considered to have failed. The test gears are spur and are immersed in 194°F oil during operation.

SYNTHETIC LUBRICANTS

Synthetic lubricants are a broad range of compounds derived from chemical synthesis rather than from the refining of petroleum. They have the following advantages:

High-temperature thermal and oxidative stability
Low-viscosity variation over a broad temperature range
Low-temperature capability
Long service life

Synthetic lubricants used in helicopter transmissions and geared gas turbine engines have the ability to operate at temperatures as low as –65°F and as high as 400°F or more during a duty cycle. These lubricants have been developed for military applications and are often designated by military specifications.

The largest class of synthetic lubricants in use today are the esters, which are materials that contain the ester chemical linkage. Two oils that are commonly used in gear applications are Mil-L-23699 and Mil-L-7808. Esters are characterized by an even balance of properties. They have wide operating

Table 5.5 AGMA Lubricant Number Recommendations for Enclosed Helical, Herringbone, Straight Bevel, Spiral Bevel, and Spur Gear Drives

Type of unit[a] and low-speed center distance	Ambient temperature[b-e]	
	−10 to +10°C (15 to 50°F)	10 to 50°C (50 to 125°F)
Parallel shaft (single reduction)		
Up to 200 mm (to 8 in.)	2–3	3–4
Over 200 mm to 500 mm (8 to 20 in.)	2–3	4–5
Over 500 mm (over 20 in.)	3–4	4–5
Parallel shaft (double reduction)		
Up to 200 mm (to 8 in.)	2–3	3–4
Over 200 mm (over 8 in.)	3–4	4–5
Parallel shaft (triple reduction)		
Up to 200 mm (to 8 in.)	2–3	3–4
Over 200 mm, to 500 mm (8 to 20 in.)	3–4	4–5
Over 500 mm (over 20 in.)	4–5	5–6
Planetary gear units (housing diameter)		
Up to 400 mm (to 16 in.) O.D.	2–3	3–4
Over 400 mm (over 16 in.) O.D.	3–4	4–5
Straight or spiral bevel gear units		
Cone distance to 300 mm (to 12 in.)	2–3	4–5
Cone distance over 300 mm (over 12 in.)	3–4	5–6
Gearmotors and shaft-mounted units	2–3	4–5
High-speed units[f]	1	2

[a]Drives incorporating overrunning clutches as backstopping devices should be referred to the gear drive manufacturer as certain types of lubricants may adversely affect clutch performance.

[b]Ranges are provided to allow for variation in operating conditions such as surface finish, temperature rise, loading, speed, etc.

[c]AGMA viscosity number recommendations listed above refer to R & O gear oils shown in Table 5.3. EP gear lubricants in the corresponding viscosity grades may be substituted where deemed necessary by the gear drive manufacturer.

[d]For ambient temperatures outside the ranges shown, consult the gear manufacturer. Some synthetic oils have been used successfully for high- or low-temperature applications.

[e]Pour point of lubricant selected should be at least 5°C (9°F) lower than the expected minimum ambient starting temperature. If the ambient starting temperature approaches the lubricant pour point, oil sump heaters may be required to facilitate starting and ensure proper lubrication.

[f]High-speed units are those operating at speeds above 3600 rpm or pitch line velocities above 25 m/sec (5000 fpm) or both. Refer to AGMA Standard 421, Practice for High Speed Helical and Herringbone Gear Units, for detailed lubrication recommendations.

Source: Ref. 2.

temperature ranges and high viscosity indexes (125 to 250). Thus esters require low torques for low-temperature operation and provide good lubrication characteristics at high temperatures.

A limitation of ester lubricants is low compatibility with some polymeric materials such as those used in seals. Also, synthetic lubricants are significantly more expensive than conventional petroleum oils.

LUBRICANT VISCOSITY SELECTION

In general, the lowest viscosity oil sufficient to form an adequate oil film at all operating conditions should be chosen. As a guide, Table 5.5 gives recommended lubricant numbers for enclosed drives. The viscosity associated with each number is presented in Table 5.2. In practice, the lubricant selection is usually a compromise between the requirements of the various oil-wetted components such as gears and bearings and the particular application requirements. For instance, a turbine generator set packager may desire a common lubrication system for the turbine, gearbox, and generator. If the turbine requires a particular lubricant such as a synthetic oil, the gear manufacturer will be requested to design for this fluid.

TYPES OF LUBRICATION SYSTEMS

There are two types of lubrication systems in use: splash and forced feed. In the splash system the unit is filled with oil to a predetermined level and operated as a sealed system with no external connections. In the forced feed system, oil is introduced into the unit through jets under pressure. Scavenge oil is pumped through a cooler and filter prior to reentering the gearbox.

The splash system is far simpler and less expensive than the forced feed design but is applicable only to low-speed units. As speeds increase, the heat generated in the gearbox becomes excessive and an external system is required to cool the lubricant. Also, oil must be precisely introduced at the gear and bearing interfaces, and this is accomplished through strategically placed jets.

For every gear drive there is a mechanical rating: the load the transmission can transmit based on stress considerations. In addition, there is a thermal rating, which is the average power that can be transmitted continuously without overheating the unit and without using special cooling. AGMA thermal ratings [3] are based on a maximum oil sump temperature of 200°F. If the thermal rating is less than the mechanical rating, additional cooling or a forced-feed lubrication system is required.

An empirical method for estimating the thermal rating of low speed (maximum rpm of 3600 or pitch line velocity of 5000 fpm) gear units is given in Ref. 3. Basically, the calculation of the thermal rating is a heat transfer problem where assumptions must be made for the heat generated in the unit and the coefficient of heat transfer of the surface area of the casing. If sufficient heat can be conducted and radiated through the casing the gearbox sump temperature will stabilize below the limiting temperature. Quite often gear units are located in areas where conditions are detrimental to good heat transfer. The following environmental conditions must be considered when defining the thermal rating [3] :

Operation in an enclosed space
Operation in a very dusty atmosphere where fine material covers the gear unit
Operation in a high-temperature ambient such as near motors, turbines, or hot
    processing equipment
Operation in high altitudes
Operation in the presence of solar or radiant energy

Auxiliary cooling can be used in combination with splash lubrication to increase the thermal rating. Air is forced past the radiating surfaces of the gear casing by strategically placed fans. The casing can also be cooled by a water jacket. In this scheme water passages are built into the gear housing, usually on the high-speed side, and heat is carried away by a cooling water flow. To operate a gear unit at maximum efficiency, auxiliary cooling schemes should include thermostatic controls so that the oil is not cooled unnecessarily. Operating with too cool a lubricant increases churning losses. The heat transfer from the gear casing can be increased by adding cooling fins which will increase the surface area.

Splash Lubrication

Successful splash lubrication of a gearbox is more a matter of development than analysis. The problem is to distribute oil to each component sufficient for lubrication and cooling yet minimize churning losses and heat generation. Figure 5.3 shows a two-stage parallel shaft unit using splash lubrication. The simplest scheme is to let the bull gear dip into the oil, as shown in the upper part of Figure 5.3, and lubricate all bearings and the high-speed mesh by the mist that is created by the action of the bull gear churning through the oil. This scheme will be more successful wtih antifriction bearings than with journal bearings, which require far more oil. To provide a positive oil supply to the bearings a feed trough can be incorporated, as shown in Figure 5.3, which catches oil flung off the bull gear. The oil drips down the casing walls, is trapped by the trough, and is distributed to the bearings.

A further degree of sophistication is shown at the bottom of Figure 5.3. An oil baffle surrounds the lower portion of the bull gear, allowing the oil level

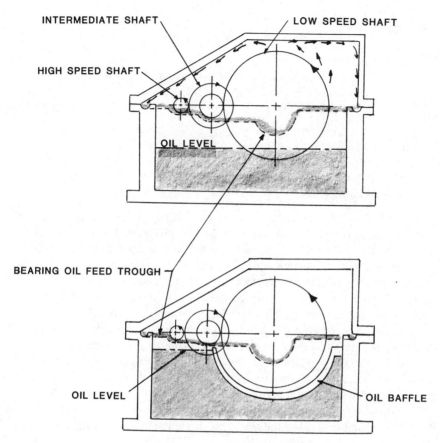

Figure 5.3 Examples of splash lubrication.

to be raised such that the large gear on the intermediate shaft dips into the lubricant and positively lubricates the high-speed mesh. Because the bull gear cannot churn through the sump, heat generation is minimized and efficiency increased. If the bull gear speed is so slow that sufficient oil is not splashed on the casing walls to lubricate the bearings, scrapers can be used to strip the oil ring that rotates with the bull gear rim. The oil is stripped from the rim and directed to the bearing feed trough (Figure 5.4). A splash lubrication system requires an oil drain, oil filler, an oil-level monitor and a breather. In cold ambients an immersion heater should be provided in the sump. An opening to the sump sufficiently large to allow cleaning of sludge after draining is also good practice.

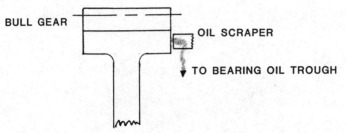

Figure 5.4 Application of oil scraper.

Forced-Feed Lubrication

Figure 5.5 illustrates a typical forced-feed lubrication system for a gearbox. The shaft-driven oil pump (A) sucks oil from the tank (D) through the suction pipe (I). From the pressure side of the oil pump the oil is directed through a cooler (C) and filter (B). A pressure relief valve (Q) is located at the inlet to the gearbox to hold the feed pressure at a constant predetermined level. An auxiliary pump (N) is incorporated to prime the system prior to starting. It could also be used as a backup in case of failure of the main pump. Check valves (G) are located such that the main pump does not pump through the auxiliary system and the auxiliary pump does not pump into the pressure side of the main oil pump. A bypass is provided at the cooler (H) which can be thermostatically controlled so that the oil is not cooled to too low a level. At various points in the system temperature and pressure sensors are located.

When designing a lubrication system the first step is to estimate the oil flow to the components and the gearbox efficiency. The temperature rise across the gearbox can then be calculated:

$$\Delta T = \frac{Q}{MC_p}$$

where

ΔT = temperature rise, $°F$
M   = oil flow, lb/min (*Note*: 1 gpm $\cong$ 7.5 lb/min)
Q   = heat generated, Btu/min [*Note*: Q = hp(42.4)]
C_p = specific heat of oil, $\cong$ 0.5 Btu/lb-$°F$

For example, a gearbox transmitting 1000 hp with 98% efficiency will reject 20 hp or 848 Btu/min of heat to the oil. If the gearbox flow is 20 GPM or 150 lb/min the temperature rise across the gearbox will be 11$°F$. In other words, if the oil enters the unit at 130$°F$ it will gain 11$°F$ and be at a 141$°F$ temperature level in the sump. Some working areas of the gearbox may be at significantly higher temperatures than the bulk sump temperature.

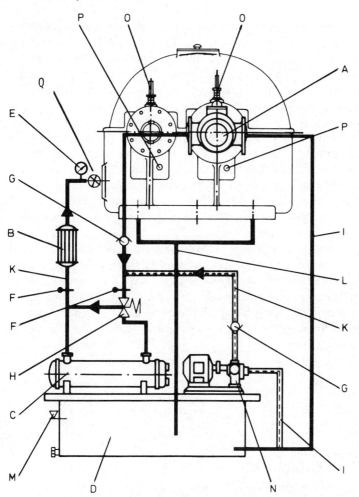

Key: A, oil pump; B, oil filter; C, oil cooler; D, oil tank; E, pressure gauge;
F, thermometer; G, check valve; H, overflow valve; I, suction pipe; K, pressure
pipe; L, return flow pipe; M, oil level gauge; N, primer and spare pump unit;
O, temperature and flow gauge; P, connection for remote thermometer;
Q, pressure regulator.

Figure 5.5 Forced-feed lubrication system. (Courtesy of American Lohmann
Corporation, Hillside, N. J.)

A typical maximum oil in temperature limit for a gearbox with forced-feed lubrication is 130°F. The cooler must be sized to cool the oil to this temperature under the worst operating conditions, such as a hot day. A typical oil temperature rise might be 30°F across the unit. These values are for mineral oils; synthetic oils can operate at higher temperature levels. Some turbine gear sets operate with oil in temperatures of 170°F.

The amount of oil flow supplied to a gear mesh is usually based on experience and experimental data. A rule of thumb might be 0.017 lb/min per horsepower or 0.002 gpm/hp. This is a generous flow and should result in a low-temperature rise. To optimize the efficiency of a gear unit a program can be conducted, gradually reducing oil flow until a predetermined maximum temperature rise is reached. Reducing oil flow will always increase efficiency, particularly in high-speed gearing, where churning is significant, but at the expense of higher scavenge temperatures.

The quantity of oil flow passing through the gearbox is regulated by the oil jet diameters, journal bearing leakages, and oil feed pressure. Usually, gearbox feed pressures are on the order of 20 to 100 psig. To hold a constant feed pressure a regulating valve is incorporated at the gearbox inlet as shown in Figure 5.5 A typical pressure regulating valve will have a spring-loaded bypass loop which opens up when the design feed pressure is exceeded and directs excess oil back to the sump.

The flow of oil through a jet can be calculated as follows:

$$Q = KA \sqrt{\frac{2G \, \Delta P}{W}} \, (W)(60)$$

where

Q = flow, lb/min
K = oil jet discharge coefficient
G = gravity acceleration, 386 in./sec^2
ΔP = pressure drop across the jet, lb/in.^{2}
W = oil specific weight, lb/in.^{3}
A = jet area, in.^{2}

If the discharge coefficient is assumed to be 0.65 and W is taken as 0.032 lb/in.^{3} the equation is

$$Q = 194A \sqrt{\Delta P} = 152(D)^2 \sqrt{\Delta P}$$

where D is the oil jet diameter inches. Note that oil flow varies with the square of the jet diameter and as the square root of the feed pressure. The minimum practical jet nozzle diameter is approximately 0.030 in. in diameter. Smaller jets tend to clog too easily due to foreign materials in the oil stream. It is good practice to have at least two jets at each lubricating position in case one clogs up.

Gears with long face widths must have oil evenly applied along the face width. If oil is introduced unevenly, thermal distortions will cause uneven load distribution. A long-face-width gear should have several jets spraying along the axis. There are also special spray nozzles that can be used to fan the oil flow out over a large area.

Relatively little oil is required for lubrication provided that it is properly applied. The bulk of the oil flow is required for cooling the gear tooth and blank. Thus it would appear logical to spray a small amount of oil at the incoming side of a gear mesh and a large amount of oil at the outgoing side, where it can do the most efficient cooling job. In practice, for low- and moderate-speed gear applications lubrication on either the incoming or outgoing side can be satisfactorily developed. For very high speed gears, above 20,000 fpm, it has been shown that lubrication on the outgoing side is more efficient than on the incoming. This is because the bulk of the oil is not being dragged through the mesh, experiencing churning losses. High-speed gear lubrication, however, is very much an art and a particular application may require extensive development to minimize thermal distortions and heat generation.

The extent to which the gear mesh is penetrated by the oil jet is a function of the velocity of the jet. The velocity in turn is a function of the feed pressure. Industrial applications generally operate with feed pressures in the order of 30 psig, whereas high-speed aerospace gearsets introduce oil at pressures of 100 psig. There are no conclusive studies determining optimum feed pressures; however, for high-speed gears it would seem that higher pressures would be beneficial.

Figure 5.5 illustrates the components incorporated in a typical forced-feed lubrication system. The following paragraphs describe in detail the major components of such a system.

PUMPS

The gearbox oil pump must deliver a given quantity of oil over a wide range of oil in temperatures and viscosities. At startup the oil will be cold and viscous. During operation on hot days the lubricant will warm up to the maximum design allowable value. Also, the pump must be capable of priming itself and overcoming the pressure drops in the line between the reservoir and the suction (inlet) port of the pump.

For gearbox lubrication rotary pumps are generally used. Figure 5.6 illustrates some widely used configurations. Rotary pumps are positive-displacement devices which deliver a given quantity of fluid with each revolution. At any particular speed the oil flow is practically constant regardless of the downstream conditions. The pressure developed is dependent on the resistance of the

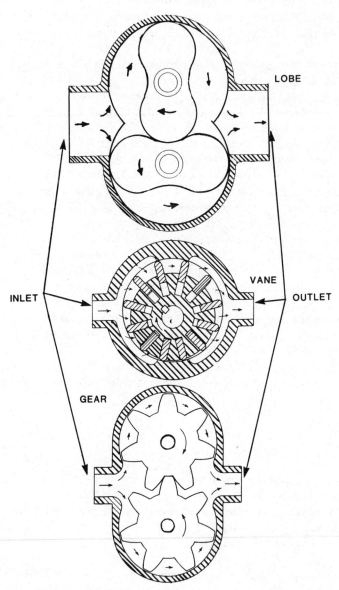

Figure 5.6 Rotary pump configurations.

discharge piping. The resistance includes the effect of oil jet orifices, pipe friction, elbows, and so on. Allowable pump discharge pressure is limited by the strength of the pump and the power of the driving unit.

Other types of pumps, such as centrifugal or piston, are not used on gearbox applications, for the following reasons:

Centrifugal pumps develop pressure as a result of centrifugal force and are used mostly where large flow volumes at relatively low pressures are required. The flow delivered varies considerably with changes in discharge pressure, and centrifugal pumps are not capable of self-priming.

Piston pumps are insensitive to discharge pressure and are self-priming; however, due to their reciprocating motion and the inertia effects of the moving parts, speeds are relatively low; therefore, pump capacities are low compared to rotary configurations. Also, the flow tends to pulsate in a piston pump configuration.

Rotary pumps in gearboxes can be flange mounted to the unit and driven by one of the gearbox shafts or independently mounted with an electric motor or other prime mover driving. When shaft driven, the flow will vary directly with shaft speed. Commercial pump speeds rarely exceed 1800 rpm in gear, vane, or lobe pump configurations, although some small aircraft types achieve much higher speeds. There are screw pump designs which can achieve speeds to approximately 5000 rpm. This is because the flow through the pump is axial and fluid speed is relatively low compared to the peripheral speeds of gear or vane pumps. Figure 5.7 is a schematic of a screw-type pump.

OIL INLET

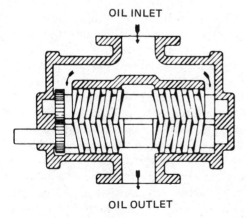

OIL OUTLET

Figure 5.7 Screw pump configuration. (Courtesy of Delaval IMP Pump Division, Trenton, N. J.)

The pump discharge pressure will depend on the required gearbox feed pressure and the pressure drop between the pump outlet and gearbox inlet. This pressure drop is a result of the losses in the filter, cooler, valves, lines, and so on, in the system. The pressure drop can be considerable and care must be taken that the pump design pressure is not exceeded. The great majority of commercial pumps are designed for pressures of 150 psi maximum, although much higher pressure capacity is possible. A pressure relief valve can be incorporated downstream of the pump outlet to protect the pump against overpressure in case of downstream blockage.

The pumping action of all rotary pumps is essentially the same. On the inlet side a void is created as the pumping elements rotate. Fluid, forced by atmospheric pressure, fills this space and is transported to the outlet side. In a gear pump, as the gears rotate, fluid is trapped between the gear teeth and the case, and is carried around to the discharge. In a vane pump, the rotating member with its sliding vanes is set off-center in the casing. The entering fluid is trapped between the vanes and the inside of the case and is carried to the outlet. The term "positive displacement pump" means that with each revolution a specific volume of fluid, depending on the geometry of the elements, is passed through the pump. If there were no clearance between the rotating elements, or between the rotating elements and the casing, the volume of fluid pumped could be easily calculated and predicted. Clearance does exist, however, and depending on the discharge pressure, there is always some internal leakage from the outlet to the inlet side of the pump. This leakage or volumetric pump efficiency must be considered when designing a pump for a specific application. Another consideration is the range of lubricant viscosity the pump will experience. Maximum internal leakage will occur when the fluid is at minimum viscosity; thus the capacity of the pump must be sized for this condition. A gearbox oil pump should be specified to be oversized at least 15% as far as flow requirement is concerned to account for operating variables and pump deterioration over time.

To control the flow into the gearbox a pressure regulating valve is usually incorporated at the unit inlet. The valve is set to maintain a constant inlet pressure, which if the gearbox orifices are correctly sized, will result in the design flow. As the oversized oil pump delivers more flow than required, the pressure will exceed the set point and the regulating valve will bypass enough flow to maintain the design pressure into the gearbox. Figure 5.8 shows the flow versus speed characteristic of a shaft driven oil pump with a pressure relief valve. From startup to point A the flow varies directly with speed. At point A the pressure, which varies with the square of the speed, has reached the set point of the relief valve and the valve cracks open. The flow into the gearbox is then regulated at a constant amount with the valve bypassing sufficient flow to keep the feed pressure constant.

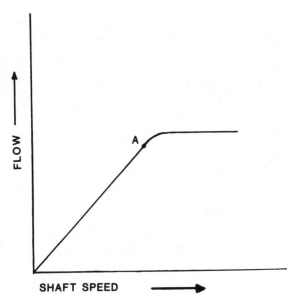

Figure 5.8 Pump flow versus speed characteristic using pressure relief valve.

A significant parameter in designing the lubrication system is the pump suction requirement. Fluid flows into a pump due to the difference in pressure between the pump inlet and fluid source. In most cases the fluid source is at atmospheric pressure, so there is 14.7 psia pressure available to force the lubricant into the pump. Tending to retard the lubricant flow are the static lift (maximum level of fluid below the pump inlet), friction losses due to pipes, valves, elbows, and so on, and entrance losses from the reservoir to the pipe and the pipe into the pump inlet. Commercial rotary pumps generate suction lifts of approximately 5 to 15 in. Hg; 30 in. Hg is equivalent to 14.7 psia; thus commercial pumps can withstand an inlet line pressure drop of approximately 2.45 to 7.35 psia. If the pressure drop in the inlet line is excessive, the pump will not fill with oil and as a result oil flow and discharge pressure will be erratic and drop off. Also, air bubbles will be formed in the fluid and as they implode, cavitation damage may occur, with the potential of pump failure.

To minimize inlet line pressure drop, the reservoir should be located as close to the pump as possible and not too far beneath it. Line and fitting sizes should be as large as possible and bends, valves, fittings, and so on, should be minimized. To help pump priming at startup, the inlet port to the pump should be full of oil. If it is possible for oil to drain down from the inlet port during shutdowns, a foot valve can be placed in the inlet line at a low point. This type of valve has a light spring load which keeps it closed at shutdown and blocks oil drainage from the pump inlet. At startup the pump suction creates a pressure differential across the valve, opening it and allowing flow. In any system when starting up after a long shutdown period it is a good idea to check the pump inlet port to ensure that it is full of oil.

A frequent problem in pump inlet systems is air leaks. Small leaks that are not easily detectable allow air into the line. The air displaces lubricant and the pump discharge pressure and flow will drop off and become erratic. It is difficult to find these leaks but they are quite common and must be considered when troubleshooting pump problems.

The power required to drive an oil pump can be estimated as follows:

$$P = \frac{Q \, \Delta p \, (231 \text{ in.}^3/\text{gal})}{(12 \text{ in./ft}) \, (33,000 \text{ ft-lb/min/hp})} = 0.0006 Q \Delta p$$

where

P = horsepower
Q = flow, gpm
Δp = pressure developed, psi

This is the theoretical power, to which must be added mechanical and viscous losses. The mechanical losses include friction drag of all moving parts, such as bearings, seals, and so on. Viscous losses include the power lost to fluid viscous drag against the pump components as well as shearing of the fluid itself. The mechanical efficiency of a pump is a measure of the magnitude of the horsepower to loss to the theoretical horsepower.

Gearboxes in critical applications often have a shaft-driven main oil pump and an electric motor-driven auxiliary pump. The auxiliary pump is used for prelubrication prior to start up and as a backup in case of failure. Suitable valving must be designed in the system (Figure 5.5) so that the pump flows are properly directed.

FILTRATION

Gearbox lubrication systems are subject to contamination due to a variety of causes. The internal components wear generating particles washed away by the oil stream. Also, at assembly, during maintenance, and even during operation foreign particles find their way into the unit, and these contaminants, if uncontrolled, will cause wear and possibly failure of bearings or gears. Filtration is the mechanism that captures particles in a fluid by passing them through a porous medium. The degree of filtration is generally specified by a micrometer rating:

$$\mu\text{m} = 10^{-4} \text{ cm} = 0.00003937 \text{ in.}$$

As a point of reference a human hair is 70 μm or 0.0028 in. thick.

Typical gearbox filtration specifications call for 40 μm or 0.0016 in. filtration, which can be interpreted to mean that the filter will allow no particle greater than 40 μm to pass. When one considers that journal bearing film thicknesses are on the order of 0.001 in. and less, and gear mesh film thicknesses lower, a case can be made for specifying finer filtration than 40 μm, possibly 10 or even 3 μm. It has been proven that fine filtration is beneficial to component life; however, to obtain full benefits great care must be taken with the entire system or the filter will clog very quickly or be in bypass. For instance, a 55-gal drum of new oil can contain a billion particles 10 μm or larger. Thus new oil should be filtered prior to use in the gear unit. Also, great care must be taken in cleaning all gearbox components and piping prior to assembly. The unit should be flushed with an auxiliary filter prior to use with the standard filter.

The absolute filtration rating of a filter is the diameter of the largest hard spherical particle that will pass through under specified test conditions. To test a filter, a measured quantity of contaminant, typically spherical glass beads in the range 2 to 80 μm diameter, are passed through the test article and captured. The captured fluid is then passed through a very fine membrane filter which is examined under a high-powered microscope. By visual observation, the diameter of the largest glass bead on the membrane is determined.

Filters are sometimes rated by efficiency. One such rating is the nominal efficiency, which is a measure of the retention by weight of a specified artificial contaminant that is passed through a test filter and captured by a membrane filter. Nominal efficiencies are typically in the range 90 to 98%; however, this rating method is losing favor due to lack of reproducibility and uniformity.

Gaining favor in the industry is the beta (β) filtration rating, which is the rating of the number of particles greater than a given size in the influent fluid to the number of particles greater than the same size in the effluent fluid passed through a test filter. For instance, if the number of particles greater than 10 μm per unit volume of fluid entering a filter is 5000 and the number of particles greater than 10 μm per unit volume of fluid leaving the filter is 50, the β_{10} rating is

$$\beta_{10} = \frac{5000}{50} = 100$$

The β rating can be converted to an efficiency as follows:

$$E_x = \frac{\beta_x - 1}{\beta_x}$$

where

$\quad E_x$ = efficiency expressed as percentage of the filter medium's ability to remove particles over a particle size x, by count

$\quad \beta_x$ = beta filtration ratio (β rating) for contaminants greater than x μm

In the previous example,

$$E_{10} = \frac{100 - 1}{100} = 99\%$$

meaning that the filter will remove a minimum of 99% of all particles greater than 10 μm. The following table gives efficiencies for various beta ratios:

β Ratios versus Efficiency

β Ratio	Efficiency (%)
1	0
2	50
20	95
50	98
100	99
1,000	99.9
10,000	99.99

Figure 5.9 shows how the beta ratio concept can be used. A β_3 filter with a ratio of 1000 will remove 99.9% of all particles greater than 18.5 μm. A β_{10} filter with a ratio of 1000 will remove 99.9% of all particles greater than 26 μm. The β_{10} filter will remove 99.7% of all particles greater than 18.5 μm.

The beta ratio is determined using a continuous flow of contaminated fluid through a test filter. Samples of the fluid from upstream and downstream of the filter are taken and analyzed with an automatic particle counter. This laboratory instrument is capable of determining the particle size and distribution per unit volume of fluid.

Filter media are usually fibrous materials comprised of many fine fibers randomly oriented with diameters ranging from 0.5 to 30 μm. Materials most commonly in use are cellulose, cotton, micro-fiberglass, and synthetics. The smaller the diameter of the fiber used, the closer they can be compacted and the higher the filter efficiency.

When designing a filter, in addition to the filtration rating, the following points must be considered:

Flow rate
Fluid type
Operating temperature
Operating pressure
Dirt-holding capacity
Pressure drop

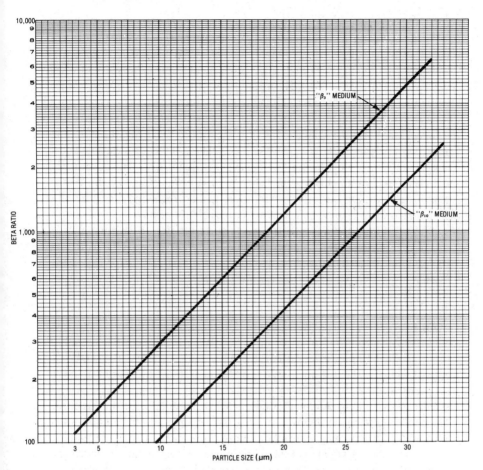

Figure 5.9 Typical beta ratios for β_3 and β_{10} depth-type media versus particle size. (From Ref. 4.)

The flow rate will determine the line size and clean pressure drop for a given size of filter. Fluid type, operating temperature, and pressure will dictate the construction and materials of the filter components. Dirt-holding capacity is an indication of how long an interval may be expected between filter changes.

As the filter collects contaminants during operation, the pressure drop across the filter will increase. Filter elements have collapse pressure ratings and the pressure drop cannot be allowed to exceed this value. Other system considerations such as exceeding the maximum oil pump discharge pressure will also affect the filter pressure drop allowed. In order to monitor the pressure differential across the filter, pressure gauges can be incorporated on the inlet and outlet sides. Filters are available with integral differential pressure indicators that can provide a warning when a set point is reached. To protect the filter and provide a continuous flow of oil, some filters incorporate a bypass valve which will open when a predetermined pressure drop is reached. At bypass, only a portion of the fluid will pass through the filter, allowing unfiltered lubricant to enter the system. For critical applications dual oil filters with changeover valves are specified. With such a design one filter can be serviced while the system is operating using the other filter. Filter elements can either be cleanable and reusable or disposable. Cleanable filter elements are usually made of wire mesh and cleaning is commonly accomplished in an ultrasonic liquid bath. For gearbox applications, in terms of filtration quality, reliability, and ease of maintenance, it is recommended to use disposable elements.

The oil filter should be located on the pressure side of the pump downstream of any component that might produce contaminants, such as the oil cooler. Filters placed in the pump suction line can cause inlet problems as the pressure drop across the filter increases with use.

COOLERS

In a forced-feed lubrication system, the oil in temperature to the gearbox is controlled by passing the hot scavenge oil through a heat exchanger. In order to specify a cooler, the maximum expected oil scavenge temperature must be estimated and the maximum allowed oil temperature into the unit defined. The cooler must be capable of achieving the required oil temperature when exposed to the maximum ambient air temperature anticipated in the application. A generous safety margin should be applied during design to account for deterioration of the cooler. For instance, over time, a water-oil cooler will experience deposits in its tubing reducing efficiency.

The two types of coolers used are liquid to liquid or liquid to air. Figure 5.10 schematically illustrates an oil-water cooler. The hot oil entering through the shell side encounters the cooling water and an equilibrium of temperatures occurs, cooling the oil and heating the water. Where water is not available, radiators are used, blowing cooling air over oil tubes. Air-to-oil coolers require larger envelopes than water-to-oil coolers. Also, on hot days, the air temperature will limit the amount of cooling a radiator can achieve.

OIL RESERVOIR

The oil reservoir or tank may be integral with the gearbox (wet sump) or separately mounted (dry sump) and connected to the gearbox by piping. If

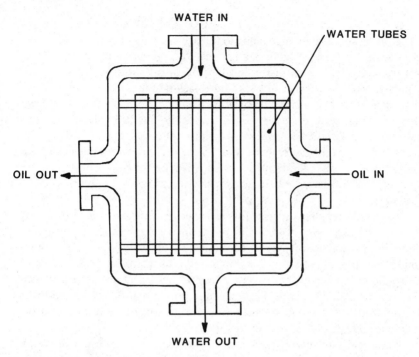

Figure 5.10 Typical oil-water cooler configuration.

integral, on high-speed gearing, the sump should be isolated from the gear windage by some type of baffling so that the oil in the reservoir is still. The level of oil in the reservoir will vary from a maximum when the unit is shutdown and oil has drained from lines and components to the minimum allowed during operation. The minimum operating level is determined by the length of time it is desired to have the oil dwell in the tank to effect deaeration. For instance, if the nominal flow is 20 gpm and a dwell time of 2 min is specified, the minimum oil volume should be 40 gal. The longer the dwell time, the more time air entrained in the oil has to settle out. It is difficult to determine how much dwell time a particular application will require by analytical means. If there is the potential for significant amounts of air to be entrained in the oil, significant dwell time may be required. Air usually enters the lubrication system through pressurized labyrinth seals. Reference 5 specifies an 8-min retention (dwell) time based on nominal flow and total volume below the minimum operating line. This is a conservative value. If there is no likelihood of significant aeration, 2 min of dwell time or less may be sufficient.

After setting the minimum operating level of the oil reservoir an estimate is made of the anticipated oil loss due to leakage and any other reasons and the time interval between oil additions in order to determine the maximum operating level to which the reservoir will be filled. At shutdown, when lines and components such as coolers and filters drain back into the reservoir, the oil level will be higher than the maximum operating level; therefore, the tank must have sufficient volume to accommodate the drain backflow and still retain some air space at the top. At initial startup some quantity of oil is required to fill all lines and components. It is good practice to run for a brief period and then check the oil level to determine if additional lubricant is required to come up to the proper operating level. Oil levels are best monitored by a sight glass and dipsticks are also used.

To ensure complete drainage for cleaning and oil changes the bottom of the reservoir should slope toward a low-point drainage connection. The oil pump suction line should connect slightly above the high end of the reservoir bottom so that sediment on the bottom is not pulled into the pump inlet line. Oil return lines should be piped into the reservoir above the maximum operating level away from the area around the pump suction connection so that the oil around this area is undisturbed. To facilitate inspection and cleaning of the reservoir, sufficiently large openings must be provided.

When cold ambient operation is anticipated, the reservoir can be heated either by applying external heat or incorporating a thermostatically controlled immersion heater in the tank. With immersion heaters care must be taken not to overheat the oil in contact with the element, since this will lead to lubricant degradation. The heater watt density should not exceed 15 W/in.^{2} [5].

BREATHER (VENT)

The gearbox breather is used to vent pressure that may be built up in the unit. Pressure may occur as a result of air entering the lubrication system through seals or the natural heating and cooling of the unit. When a cold gearbox starts up, the heat generated during operation causes air within the case to expand. The same effect is noted with units operating outdoors as they heat up during the day and cool at night.

Breather caps usually have some type of baffling built in to keep particulate contamination out of the gearbox, but moisture can enter. It is possible to have a completely sealed system and incorporate an expansion chamber in place of the breather. Figure 5.11 presents a schematic of an expansion chamber which is screwed into the gear case or reservoir. The expansion or contraction of air is accommodated by flexing of the diaphragm. In harsh environments care must be taken in choosing the diaphragm material such that it is compatible with the environment.

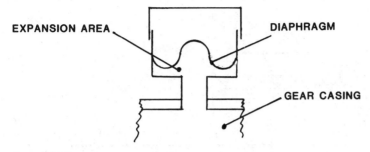

Figure 5.11 Expansion chamber.

In a complicated system the gearbox vent may be inadvertently routed to an area with an unfavorable environment and cause operating difficulties. For instance, if the gearbox is vented to a pressurized area, the gearbox may be back-pressured, resulting in oil leakage through the seals. If vented to an area under a vacuum, oil might be sucked out of the unit through the vent.

PIPING

Although apparently simple, the piping connections for a gearbox can be a source of aggravation at assembly and startup. In many cases the responsibility for supplying piping and lubrication system components is split between the gear manufacturer and the user. The purchase order should be specific as to who supplies what, which specifications apply, and where the interfaces are in order to avoid complications at installation.

As far as gearbox oil feeds are concerned, it is good practice to have only one external connection with all other oil passages placed inside the casing. Then if there is any slight leakage in the piping connections, it will be internal and harmless. Also, there will be less chance of damage to the piping during shipping and assembly. When units are horizontally split, the lubrication piping and components should be mounted on the bottom of the casing such that the top half can be disassembled without disconnecting any lubrication lines.

The piping arrangement must be carefully designed to minimize pressure drops. Care must be taken to position components such as coolers and filters such that they will not drain at shutdown and require refilling at each startup. There are several methods of fabricating lubrication system piping. Reference 5 specifies bending the pipes to suit the required contours and welding them at connections. This is the best way to ensure that there will be no leakage; however, it is more expensive and requires more skill than using threaded joints. The piping terminates in flanged connections which are bolted to the mating part. Seamless carbon steel piping to ASTM A106 or ASTM192 is recommended.

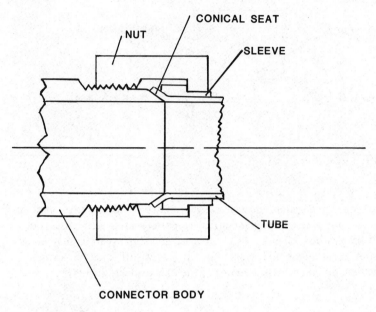

Figure 5.12   Flared fitting.

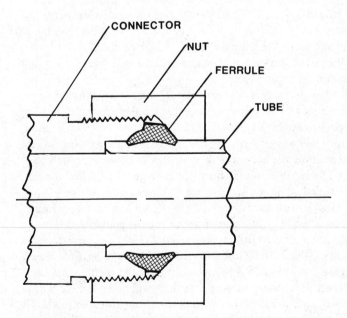

Figure 5.13   Flareless tube fitting.

There are two basic types of threaded joints. One is the pipe thread, which is tapered and produces a metal-to-metal seal by wedging surfaces together as the pipe is screwed in. When connecting this type of joint, sealant should be applied to the male thread. Avoid the first two male threads from the end to keep sealer out of the system. Tetrafluoroethylene-fluorocarbon tape is not recommended to seal pipe threads since pieces may break off and enter the lubrication system, clogging the oil jets. SAE straight thread fittings depend on an O-ring for sealing. Care must be taken in assembly or the O-ring will be damaged, resulting in oil leakage. More positive in stopping leakage than threaded fittings are tube fittings, which fall into three groups: flared, flareless, and the previously mentioned welded or brazed. A typical flared fitting consists of three pieces (Figure 5.12). The seal occurs when the soft tubing is pressed against the hardened conical seat of the connector body. Standard flare angles are 37° and 45° and should never be mixed.

Flareless fittings are usually used where tube thicknesses make use of flared tubes difficult. Figure 5.13 illustrates a bite-style flareless tube fitting where the wedging action of the ferrule provides the seal.

REFERENCES

1. E. R. Booser and R. C. Elwell, Keeping Lubricants Flowing at Low Temperatures, *Machine Design*, March 24, 1977, pp. 74–79.
2. AGMA Standard 250.04, Lubrication of Industrial Enclosed Gear Drives, American Gear Manufacturers Association, Arlington, Va., 1981.
3. AGMA Standard 420.04, Practice for Enclosed Speed Reducers or Increasers Using Spur, Helical, Herringbone and Spiral Bevel Gears, American Gear Manufacturers Association, Arlington, Va., December 1975.
4. Filtration Manual, Purolator Technologies, Newbury Park, Calif., 1979.
5. API Standard 614, Lubrication, Shaft Sealing and Control Oil Systems for Special Purpose Applications, American Petroleum Institute, Washington, D.C., September 1973.

6

MATERIALS AND HEAT TREATMENTS

The selection of the gear material and its heat treatment is the most basic decision in the design process. The strength of a gear tooth, as shown in Chapter 3 is proportional to its hardness; therefore, the size, configuration, and cost of a gear unit are dependent on the choice of gear material and its processing. Although there are a wide variety of materials and processes to choose from, practical considerations tend to narrow the options. By this is meant that at any gear manufacturer's facility the type of equipment available for tooth finishing and heat treating will limit the designer's choices. In fact, the equipment available dictates the design philosophy and the material and processing specifications.

The types of gearboxes discussed in this book usually use gears made of alloy steels. These materials carry the greatest load in terms of power transmitted per pound of gear and offer high reliability. Steel is defined as iron with carbon percentages in the range of 0.15 to 1.5. A steel is considered to be an alloy when the maximum of the range given for the content of alloying elements exceeds one or more of these limits; 1.65% Mn, 0.60% Si, or 0.60% Cu; or when a definite range or minimum amount of any of the following elements is specified: aluminum, chromium, cobalt, columbium, molybdenum, nickel, titanium, tungsten, vanadium, or zirconium. Table 6.1 lists the chemical compositions of some of the most widely used gear steels. The American Iron and Steel Institute (AISI) or Society of Automotive Engineers (SAE) numbers designate the steel composition and alloy type. The last two digits signify the carbon content: for example, AISI 4140 contains 0.40% carbon. The first two digits designate the approximate alloy content.

Gears are generally either through-hardened in the range Rc 32 to 43 (Bhn 300 to 400) or surface hardened in the range Rc 55 to 70. The hardness

Table 6.1 Chemical Compositions of Gear Steels

AISI number	C	Mn	Ni	Cr	Mo
4140	0.38/0.43	0.75/1.0		0.80/1.0	0.15/0.25
4340	0.38/0.43	0.60/0.80	1.65/2	0.7/0.9	0.20/0.30
4620	0.17/0.22	0.45/0.65	1.65/2		0.20/0.30
4320	0.17/0.22	0.45/0.65	1.65/2	0.4/0.6	0.20/0.30
8620	0.18/0.23	0.70/0.90	0.40/0.70	0.4/0.6	0.15/0.25
3310	0.08/0.13	0.45/0.60	3.25/3.75	1.4/1.75	
9310	0.08/0.13	0.45/0.65	3/3.5	1/1.4	0.08/0.15
2317	0.15/0.20	0.40/0.60	3.25/3.75		

range between Rc 43 and Rc 55 is seldom used since the steel is too hard to cut in this state. Therefore, it may as well be fully hardened to obtain maximum load capacity and finish ground if necessary.

The strongest and most durable gear meshes are made up of two surface-hardened gears using the carburizing process. Utilization of surface-hardened gears will result in the minimum gearbox size, creating savings in materials, machining, and handling costs. On the other hand, there are increased costs associated with carburized gears. Because of heat-treat distortion, griding after hardening will be necessary to achieve high precision of the gear teeth. The grinding and heat treating equipment is very costly and may not be readily available. From a technical point of view, however, carburized, hardened, and ground gears are the best alternative for the high-speed, high-power drives. Of the materials listed in Table 6.1, the low-carbon steels—4620, 4320, 8620, 3310, 9310, and 2310—are carburizing grades.

Another surface-hardening process that is used extensively is nitriding. This process produces a very hard, wear-resistant case which is somewhat more brittle than that of a carburized gear; thus the fatigue properties are not as good. Heat-treat distortion resulting from nitriding is not as great as that from carburizing; thus it is sometimes possible to use gears after nitriding without a grinding operation. Even if grinding is planned, nitriding is sometimes used to limit distortion. It is not uncommon to mesh a carburized and ground pinion with a nitrided gear. When the gear is large, grinding is often considered impractical either because equipment is not available or simply because the cost is too high. AISI 4340 and 4140 are nitriding steels.

Another common material combination is a surface-hardened pinion meshing with a through-hardened gear. Although the gear strength and durability is considerably lower than that of the pinion, the gear will experience fewer stress cycles and a reasonable fatigue life can be achieved. The hardened pinion may tend to wear the softer gear in improving mesh characteristics and also cold working the gear, which increases its surface hardness. The through-hardened gear is considerably cheaper than a surface-hardened gear since the heat treatment is much simpler and no grinding is required.

The least expensive mesh combination is a through-hardened pinion mating with a through-hardened gear. The size of such a unit will be large compared to a gearbox incorporating surface-hardened gears: as much as twice the envelope and weight; however, where there are no restrictions on size or inertia, this type of design is widely used. Most double helical units have through-hardened gears. Of the materials listed in Table 6.1, AISI 4140 and 4340 are through-hardening steels.

There are other methods of surface hardening in addition to those mentioned above, such as induction or flame hardening. Also, other materials are in use, such as AISI 1040, 1045, 1137, and 1340. These are plain carbon steels where the alloying element content is low. These materials and processes are economical and satisfactory for gears requiring only a moderate degree of strength and impact resistance. Where cost is the primary consideration, these options should be considered.

HARDENING FUNDAMENTALS

Steel, as received from the mill, has a structure of nearly pure iron (ferrite) and iron carbide (cementite). To harden a part it has to be heated in the range of 1450 to 1600°F, where it takes the form of austenite, a structure of iron and carbon stable at high temperatures. Depending on how quickly the part is cooled, the austenite transforms to other structures. Figure 6.1 illustrates the formation of the various structures of carbon steel due to quenching. If cooled very rapidly, a hard, strong, and somewhat brittle form called martensite, which is carbon dissolved in iron, is formed.

The maximum hardness that any steel can attain after quenching is a function of its carbon content. Figure 6.2 illustrates this point. The curve reflects 100% transformation of austenite to martensite. From the curve it can be seen that the through-hardening, high-carbon steels 4140 and 4340 can attain a maximum hardness of Rc 58 after quenching. Gears are not used in this state since a fully quenched martensitic structure lacks the touchness to be able to resist shock loading. To improve the gear tooth's ductility and toughness its structure is modified by tempering, a controlled reheating of the part to a temperature below the austenitizing temperature. This process also reduces residual stresses left in the gear during hardening. Tempering temperatures as low as 300°F can produce significant increases in toughness and ductility and reductions in residual stresses with small hardness decreases.

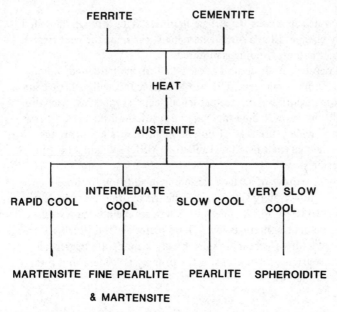

Figure 6.1   Formation of steel structures during quenching.

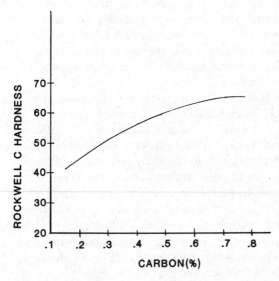

Figure 6.2   Relationship of carbon content to hardness.

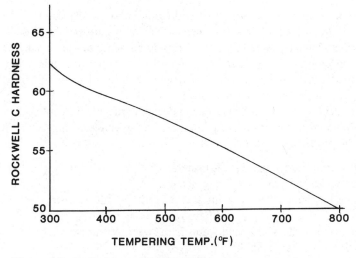

Figure 6.3 Reduction of steel hardness due to tempering.

In the case of carburizing steels which initially have low carbon levels, the tooth surface is enriched with carbon to approximately 0.85% and thus can attain a surface hardness of approximately Rc 65 maximum. The core will remain relatively soft, about Rc 35; thus the carburized tooth has a strong hard case with a tough ductile core. This structure has been found to be optimum for load transmisision. The carburized gear, like the through-hardened gear, must be tempered after hardening. Figure 6.3 shows the surface hardness drop-off of an AISI 9310 carburized gear at various tempering temperatures. The surface hardness was Rc 65 in the as-quenched condition.

HARDENABILITY AND THE USE OF ALLOY STEELS

An alloy steel is chosen on the basis of the mechanical properties required in the part and the heat treatment available to develop these properties. The characteristic of a steel that will determine if it is acceptable is called hardenability. Hardenability is a measure of the severity of cooling conditions necessary to achieve the hardness required. Hardenability of an alloy steel is established by the carbon and alloy content. To obtain a given hardness, higher-hardenability steels require less severe quenching than do low-hardenability steels. A less severe quench will result in lower distortion of the part and less chance of residual stresses, which can manifest themselves in cracks during processing or gear operation.

   For the same section thickness a steel with good hardenability can be hardened throughout with a far less severe quench than a low-hardenability steel. For instance, a 1-in.-diameter AISI 1045 steel bar fully hardened at the outside when quenched in water will have a hardness drop-off at its center of 5 points Rc. A 1 in. diameter AISI 4340 bar will have negligible hardness drop-off through its cross section after water quenching.

   In order to achieve lower cooling rates and deeper hardening properties, alloys are incorporated in the steel formulation. The following elements are added to alter the properties of steel:

Carbon. The principle hardening element in steel
Manganese. A strengthening and hardening element
Nickel. Improves low-temperature toughness and ductility, reduces distortion
         and cracking during quenching, and improves corrosion resistance
Chromium. Increases hardenability and forms hard, wear-resistant carbides
Molybdenum. Improves hardenability

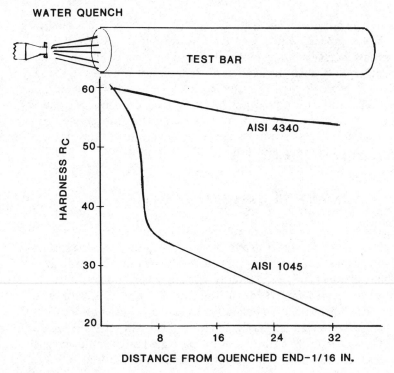

Figure 6.4 Jominy test for hardenability.

The standard method of determining a steel's hardenability is the Jominy bar test. A vertically suspended bar 1 in. in diameter and 4 in. long is heated above its critical temperature into the austenitizing range. It is then water quenched at one end and air cooled at the other end. Hardness is measured along the bar and the gradient between the fully hardened water-quenched end and the softer air-cooled end represents the hardenability of the steel. If the steel has high hardenability, the hardness will be at a higher level farther from the water-quenched end than that of a low-hardenability steel. Figure 6.4 illustrates the results of a Jominy bar test.

STEEL QUALITY

The AISI designation of a steel specifies only the chemical composition of the material. The quality in terms of freedom from seams, cracks, folds, inclusions, or nonhomogenous structure may not be adequately controlled when specifying a material in this manner. When the best quality is required, steels should be produced by the electric furnace process rather than the basic open hearth or oxygen process. Macroscopic, microscopic, and magnetic particle inspection should be required to determine material structure and cleanleness. For critical, high-reliability applications it may be necessary to specify that material be produced by vacuum or double vacuum melting to assure adequate material cleanliness.

The Society of Automotive Engineers issues a series of Aerospace Materia! Specifications (AMS) that control the quality of steels. For instance, AMS 6260

Table 6.2 AMS Specifications for Gear Steels

AISI (SAE) designation	AMS designation	Remarks
4140	6382H	
4340	6414B	Premium quality, cons. elec. vacuum melted
4340	6414H	
4620	6294D	
4320	6299B	
8620	6276D	Premium quality, cons. elec. vacuum melted
8620	6277B	Premium quality, cons. elec. melted
3310	6250G	
9310	6260H	
9310	6265D	Premium quality, cons. elec. vacuum melted
9310	6267B	Premium quality
2317	—	

covers air-melt AISI 9310 steel and AMS 6265 is the vacuum-melt version. The AMS numbers corresponding to the steels in Table 6.1 are given in Table 6.2.

The steel melting process is critical in controlling material cleanliness. Air melt is the standard procedure. Premium-quality steels undergo vacuum degassing where the billet is heated to a temperature slightly below the melting point in a vacuum environment and hydrogen and oxygen are removed from the steel. This process minimizes nonmetallic inclusions such as aluminum oxide and silicon oxide in the material. A step further in producing clean steel is to melt the material one or more times in a vacuum so that impurities are removed as gases. It has been shown that vacuum melt steels yield significant improvement in gear reliability by minimizing the possibility of fatigue crack initiation due to material impurities. Specifying an AMS steel may add to the cost of a gear initially but in many instances can save the manufacturer the cost of scrapping a semifinished part when material defects are uncovered during machining operations.

PROCESSING OF THROUGH-HARDENED GEARS

Steel, as obtained from the mill, either in an as-forged or rolled condition may not be uniform with high- and low-hardness areas. The first metallurgical operation to be performed is annealing, which involves slow heating, austenization of the steel, and then slow furnace cooling. Annealing will improve machininability by providing a uniform low-hardness structure. This facilitates the initial rough machining.

The part may then be normalized, which involves heating above the austenitizing temperature and then air cooling at an intermediate rate. The resulting structure is somewhat harder, stronger, and less ductile than when in the annealed state. The gear is finish machined at this point except for the teeth and bearing journals. The normalized structure facilitates precise machining and results in a good surface finish.

The part is then hardened and tempered to the design requirement. At this point the gear teeth are cut and the bearing journals finished. Another temper is performed to relieve machining stresses and possibly a finishing operation is required on the teeth such as shaving or lapping. The journals might also require refinishing.

CARBURIZING

In the carburizing process carbon is diffused into the surface of the teeth by controlled exposure at temperatures of $1650°F$ or above for the length of time necessary to achieve the desired case depth. After hardening the tooth has a high surface hardness, decreasing to a specified core hardness which depends on the carbon and alloy content of the particular steel used. Carburizing produces the strongest tooth with respect to bending and pitting resistance by generating residual compressive stresses in the case area. The hard surface has excellent wear and scoring resistance.

Before carburizing, the gear teeth are cut and other areas that are to be hardened are machined. Areas of the part that will remain soft must be insulated from the carburizing medium. This may be accomplished by copper plating or other masking procedures or by leaving excess stock to be machined after heat treating. In some cases a carburized area can be machined after hardening if it is to remain soft. After carburizing the part is hardened by quenching. The process will cause distortion in the part; therefore, to attain precise tooth geometry the gear teeth must be ground after hardening. It is important to minimize the distortion so that the stock removal after hardening is small since the grinding process is removing the hard case that is desired. In many cases the gear must be quenched in a die so that distortion can be kept within acceptable limits.

During the carburizing process carbon is introduced in liquid or gaseous form. The most common process now in use is gas carburizing, where a controlled gas surrounds the part in a sealed furnace. The amount of time required to carburize a part is dependent on the temperature; however, carburizing above $1800°F$ tends to coarsen the grain size to an unacceptable level.

The optimum carbon content to achieve maximum hardness is 0.080 to 0.090%. This is achieved by controlling the richness of the carbon medium. The case depth is controlled by taking sample coupons out of the furnace at intervals during the cycle. When the coupon achieves the required case depth the carburizing cycle is stopped.

Table 6.3 Recommended Case Depth versus Diametral Pitch for Carburized Gears

Diametral pitch	Light case (in.)	Standard case (in.)	Heavy case (in.)
4	0.041	0.053	0.064
6	0.030	0.040	0.049
8	0.028	0.032	0.040
10	0.020	0.027	0.034
20	0.010	0.014	0.018

Source: Ref. 1.

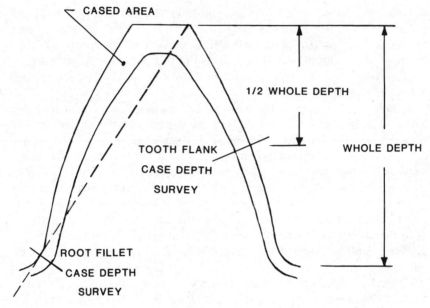

Figure 6.5 Case depth measurement.

The effective case depth is defined as the depth at which the hardness of the case is Rc 50. The case depth must be sufficiently deep to withstand the stresses created by the transmitted load, yet not so deep that the tip of the tooth becomes brittle with a tendency to break off. For this reason it is recommended that tooth tips and end faces be masked off during carburizing and left at core hardness. The case depth required will vary according to the tooth size or diametral pitch. Table 6.3 gives recommended case depths for various diametral pitches.

The case depth is measured by microhardness-testing a cross section of a tooth. This requires sacrificing a production piece or including a test section of gear teeth with the heat-treated lot. The case depth measurement is accomplished in the laboratory using an instrument which measures the hardness from the surface inward every 0.001 in. The hardness survey should be made at the tooth midheight perpendicular to the tooth surface. The root fillet region should also be checked since it is harder for the carbon to penetrate this closed-in area. Figure 6.5 illustrates the areas of the tooth where the survey should be taken. It is not uncommon for the root hardness and case depth to be somewhat lower than the flank, but both should be within design requirements.

In addition to the case depth, the engineering drawing should specify the following:

Minimum case hardness or range. The maximum carburizing surface hardness attainable after grinding is approximately Rc 63; therefore, the most stringent practical drawing requirement is Rc 60 to 63. In many cases Rc 58 minimum is specified and for moderately loaded gears Rc 55 minimum may be acceptable.

Core hardness range. Core hardness should be held within the range Rc 30 to 40. Higher core hardnesses may make the tooth too brittle, and lower values have insufficient strength. It is difficult to control core hardness closely since this parameter is dependent on the base steel carbon content, which varies from one heat value to another. The process controls are directed toward achieving case hardness and the core hardness falls where it will.

There are variations possible in the process used to carburize gears which involve a trade-off between cost and the quality of the metallurgical structure obtained. Table 6.4 shows a carburizing process for a critical 30-tooth, 10-pitch gear with a 1.0-in. face width. This is an elaborate process designed to achieve the best possible metallurgical structure. Note that following carburizing there is a subcritical annealing step the purpose of which is to refine the case structure. This step is sometimes omitted and the part is cooled down to 1500°F and quenched directly from the carburizing furnace. If this is done, care must be taken during the cooling to surround the gear with a carbon atmosphere so that decarburization of the part does not occur. A similar problem arises in the subcritical annealing step. If an air atmosphere is used, the part must be copper plated so that scale formation does not occur during annealing or decarburization during the subsequent hardening.

The hardening process involves heating the gear until the part is completely austenitized and quenching in an oil bath at 80 to 140°F. The objective of the quench is to transform the austenite into martensite; however, in practice, all the austenite can never be transformed. This is undesirable since the retained austenite is not stable and if present at too high a level can spontaneously cause distortion in the part. In order to transform as much of the retained austenite as possible, carburized parts are sometimes subzero cooled in the range -100° to -150°F. This cooling must be accomplished within 20 min of quenching, to avoid austenite stabilization. The minimum volume of retained austenite that can be practically achieved is 10%. For applications that are not extremely critical, 20% can be allowed. Retained austenite is measured definitively by x-ray diffraction or (visually estimated by) metallographic examination of an etched cross section.

Carbides in the case structure are desirable for increased strength and wear resistance, however, if they form continuous or semicontinuous networks at the grain boundaries, stress concentrations result which can initiate crack.

Table 6.4 Carburizing Process for AISI 9310 Steel

Process operation	Temperature (°F)	Time at temperature (hr)	Type atmosphere	Type of Quench	Quench temperature (°F)	Remarks
Normalize	1725	2	ENDO[a]	Chamber	R.T.[b]	
Harden	1500	1	ENDO	Oil	130	5 min in oil
Temper	1050	4	Air	Air	R.T.	
Carburize	1700	As required	Carbon potential	—	—	Furnace cool to 1500°F, air cool to R.T.
Subcritical anneal	1150	2	Air			
Harden	1500	1	ENDO	Oil	130	Agitated oil
Freeze	−120	1		Liquid		Thaw to R.T.
Temper	300	4		Air		

[a]Endothermic atmosphere.
[b]R.T., room temperature.
Source: Ref. 1.

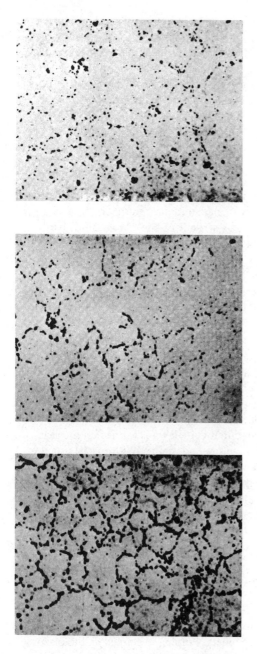

Figure 6.6 Typical metallographic standards for undissolved case carbides. (From Ref. 1.)

Figure 6.6 illustrates the levels of severity of grain boundary carbides. These are photographs of an etched cross section at 500X magnification. Figure 6.6A shows scattered grain boundary carbides and is the maximum condition acceptable for critical applications. Figure 6.6B illustrates a semicontinuous grain boundary carbide network. This is a marginal condition and might be acceptable for noncritical gearing. Figure 6.6C is a continuous grain boundary carbide network and is not acceptable.

NITRIDING

Gas nitriding is a surface-hardening process in which the part is heated to approximately $1000°F$ in a furnace with an ammonia atmosphere. Quenching is not required, the hardness developing as a result of the formation of hard nitrides near the surface. AISI 4340 and 4140 are nitriding steels, their chromium and molybdenum alloys combining with the nitrogen to form nitrides. The steel must be hardened and tempered before nitriding to a tempering temperature higher than the subsequent nitriding temperature. The nitriding cycle is quite long. For instance, a 0.020-in. case depth requires approximately 40 hr of processing.

Because the part is not quenched and the process is performed at a relatively low temperature, little distortion occurs during nitriding. For this reason many gears are finished by cutting, shaving, or grinding in the soft state and then hardened by nitriding. For extreme accuracy grinding after nitriding is still required; however, for some large gears the process is still preferred to carburizing because of the limited distortion.

During the nitriding process an outer white layer is formed on the gear teeth which is hard and brittle and therefore undesirable. With special processing the white layer can be held to a maximum depth of 0.0005 in. which is usually acceptable. Thicker white layers are unacceptable because of the possibility of spalling or flaking. They must be removed either by grinding, some other mechanical process or chemical action. Nitriding produces shallow case depths compared to carburizing; therefore, if grinding after nitriding is required, stock removal must be closely controlled to preclude grinding away the case.

FORGING OF GEAR BLANKS

Gears up to approximately 6 in. in diameter can be fabricated from bar stock. Larger gears are forged from bars or billets. Forging is defined as the plastic deformation of a metal at an elevated temperature into a predetermined size or shape using compressive forces exerted through some type of die by a

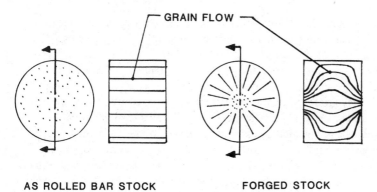

Figure 6.7 Grain flow due to forging.

hammer, press, or an upsetting machine. Many gear blanks are made of upset forgings, where a round bar is struck axially between two flat dies. The bar is shortened and the metal, following the path of least resistance, flows outward.

A forged part will have flow lines which are composed of fibrous nonmetallic inclusions or segregated phases which have been flowed in the direction of working. These fibers, if properly oriented, add to the strength of a part. Figure 6.7 shows the grain glow in a part made from bar stock compared to the same part made by upset forging. The direction of the forged flow lines will improve the bending resistance of the gear that will be fabricated from the bar.

HOUSING MATERIALS

In the design phase of a gearbox the question arises s to whether the housing should be cast or fabricated. A fabrication is comprised of steel plates which are welded or bolted together to form the casing. Usually, a carbon steel such as AISI 1020 is used.

If only one or two units are to be manufactured, the economics favor fabrication. In order to use a casting a pattern must be developed and the cost of fabricating a pattern is significant. Also, if the program is under time pressure, a fabrication can be made in less time than required to develop a pattern and pour a casting.

As the quantity of gearboxes required increases, it becomes more economical to go to a casting. The pattern costs are amortized over the production quantity and the basic material cost of cast iron is lower than steel. Also, less material is used since the casting can be shaped to conform closely to the gear configuration.

There are technical advantages to a casting in that the gearbox weight can be reduced and the casting provides more flexibility to the designer to achieve noise and vibration attenuation.

Cast irons are alloys of iron and carbon containing more than 2% carbon. The types of cast irons most commonly used for gear casings are gray iron and ductile iron. They are specified by American Society of Testing and Materials (ASTM) standards, which define chemical composition and physical properties.

Gray cast iron is graded by minimum tensile strength. For instance, a class 30 gray iron would have a minimum tensile strength of 30,000 psi. Gray iron grades range from 20 to 60. Ductile iron is specified by a three-part system. For instance, an 80-55-06 alloy has a minimum tensile strength of 80,000 psi, a minimum yield strength of 55,000 psi, and 6% elongation in 2 in. Ductile iron classes range in minimum tensile strength from 60,000 to 120,000 psi.

The type of cast iron selected for a given application must take into account not only the physical properties desired but the size and shape of the casting. The solidification rate, which depends on casting geometry and foundry practice, profoundly influences the resulting strength. Actual properties of a cast part will vary with the cooling rate, cross section, microstructure, and the grade of iron used. Like steel parts, cast irons are heat treated for a variety of reasons. These reasons can be summarized under the headings of stress relief, annealing and normalizing, and hardening and tempering.

The stress relief treatment is used because cast irons are susceptible to growth when allowed to stand at room temperature for long periods of time. Stress relief is accomplished at temperatures of 700 to 1300°F. If time permits, castings can be aged for 6 months or more to allow growth to occur and the stress relief can be omitted.

Annealing and normalizing heat treatments are used to modify the castings metallurgical structure to increase machinability. Hardening and tempering, as in steel parts, are applied to improve component strength.

In the casting process discrepancies sometimes occur which leave voids in the material. These can sometimes be repaired by welding. Welding reworks should be inspected by x-ray techniques to ensure a completely crack-free repair.

HARDNESS TESTING

Because so much attention is paid to gear tooth hardness it is appropriate to review the definition of hardness and how this material property is measured. Hardness of a material can be defined as its resistance to permanent deformation. There are several methods to measure this property, the most widely used being Brinell and Rockwell.

In the Brinell test, a load, usually 3000 kg, is applied to the test piece by means of a hardened 10 mm steel or tungsten carbide ball. The resulting impression is measured and the Brinell hardness number (kg/mm^2) is calculated by dividing the load applied by the area of the impression. The Rockwell hardness test uses a diamond cone to indent hard materials. First a minor load of 10 kg is applied to force the penetrator below the surface of the material, following which a major load varying between 60 and 150 kg depending on the scale used is applied. The hardness, which is proportional to the depth of penetration, is read directly. The Rockwell machine is more flexible and easier to use than the Brinell method; therefore, it is widely used for routine testing and the inspection of heat-treated parts. There are two types of Rockwell machines, the Normal tester for relatively thick sections, and the Superficial tester for materials too thin to be tested with the normal instrument. Minor loads on the superficial tester are 3 kg and the major loads vary from 15 to 45 kg. A cone-shaped diamond indenter is used. For work with gear steels the A, C, and D scales on the Normal tester are used. On the Superficial tester the 15-N and 30-N scales are used.

In the laboratory microhardness testing is accomplished using two different types of diamond indenters. One is a pyramid with a square base [diamond pyramid hardness (DPH)] and the other is a rhombic-based pyramid

Table 6.5 Approximate Comparison of Hardness Scales

Brinell 10-mm ball, 3000-kg load	Rockwell			DPH	Knoop
	15-N (15-kg load)	C (150-kg load)	30-N (30-kg load)		
614	90	60	77.5	695	732
587	89.25	58	75.5	655	690
560	88.5	56	74	617	650
534	87.5	54	72	580	612
509	86.5	52	70.5	545	576
484	85.5	50	68.5	513	542
460	84.5	48	66.5	485	510
437	83.5	46	65	458	480
415	82.5	44	63	435	452
393	81.5	42	61.5	413	426
372	80.5	40	59.5	393	402
352	79.5	38	57.5	373	380
332	78.5	36	56	353	360
313	77	34	54	334	342
297	76	32	52	317	326
283	75	30	50.5	301	311

(Knoop). Loads applied are on the order of 5 to 100 kg and the hardness number is proportional to the ratio of the load to a characteristic length of the impression to the second power. The test specimen is an etched, mounted cross section of tooth and the dimensions of the impression left by the indenter are read with a microscope. The DPH numbers correlate well with Knoop results up to a hardness of approximately Rc 58. Beyond this level a Knoop reading of a given specimen referred to the Rockwell C scale will yield a value 1 or 2 points lower than a DPH reading of the same specimen. Table 6.5 gives approximate hardness comparisons for Brinell, Rockwell, DPH, and Knoop. Hardness tests are usually not performed on tooth flanks since the indentation can be a source of stress concentration. On some designs there may be hardened areas which are not functional and can be surfaces for hardness checking. In many cases the gear tooth hardness is verified only by the microhardness survey of the cross-sectioned tooth sample taken from the heat-treated lot. There is nothing wrong with this practice and it is the best way to confirm hardness and case depth in the tooth root.

NONDESTRUCTIVE TESTING

Nondestructive tests are used to detect mechanical defects in material and variations in condition or composition. These tests are performed on production parts and do not impair the component's function.

MAGNETIC PARTICLE INSPECTION

This inspection method is used to identify discontinuities such as cracks, inclusions, or pores which are at or near the surface of a ferromagnetic material. Such discontinuities are areas of stress concentration and can propogate under cyclic loading leading to component failure. Cracks can be a result of poor grinding process or heat treating; therefore, it is good practice to provide for magnetic particle inspection at various stages in the manufacturing cycle. When a defect occurs it is best to identify it as early as possible so that minimum machining time is spent on discrepant material.

Magnetic particle inspection is accomplished by magnetizing a part by passing an electric current through it. Leakage magnetic fields occur at surface or near surface discontinuities. These defects are revealed by their attraction for finely divided magnetic particles which are introduced on the surface of the part being inspected.

The widely used Magnaflux method applies iron oxide particles suspended in kerosene or oil just before or while the part is being magnetized. This is called

Figure 6.8   Magnetic particle indications (indicated by arrows).

the wet method. There is a dry method where powdered iron oxide is applied while the part is under magnetization. The iron oxide particles cluster about the discontinuity and reveal its approximate location, size, and extent. The black particles are sometimes colored red or gray for better contrast on dark objects. There is a Magnaglo method of inspection where the particles are coated with a fluorescent material which glows when activated by ultraviolet light. This improves the contrast between the iron oxide particles and the part under investigation.

It is important to demagnetize any part that has undergone magnetic particle inspection. If gear teeth or shafts have any residual magnetism, they will tend to attract machining chips or abrasive particles, which can cause difficulty during operation.

To determine whether a part is acceptable following magnetic particle inspection, acceptance and rejection standards must be defined. In general, no indications of discontinuities are allowed on critical areas such as gear or spline teeth. Also, indications that extend over or into an edge, chamfer, corner, radius, fillet, or hole are not acceptable. Often, magnetic particle indications can be removed by localized grinding or remachining. For instance, indications at the edge of a tooth might be removed by regrinding a larger edge radius.

In noncritical areas some indications may be accepted. A typical specification may be as follows:

A maximum of three indications 3/16 in. long or less are allowed per noncritical area.

Figure 6.8 presents some examples of magnetic particle indications.

LIQUID PENETRANT INSPECTION

Liquid penetrant inspection is a nondestructive method for finding discontinuities that are open to the surface of solid and essentially nonpourous materials. In gearboxes the process is usually used for the inspection of castings. The principle of the liquid penetrant inspection method is to wet the surface of the workpiece with a uniform liquid coating which migrates into cavities that are open to the surface.

A widely used method is the Zyglo inspection, which uses nonmagnetic particles that fluoresce under ultraviolet light. The particles are suspended in oil and applied to the part to be inspected. Excess liquid is removed from the part and the material remaining in any cracks or discontinuities is revealed under inspection by ultraviolet light.

It is difficult to set standards of acceptance for discontinuities found in castings since the shapes are usually unique to the particular application. The location of the defect must be taken into account, and whether it may be

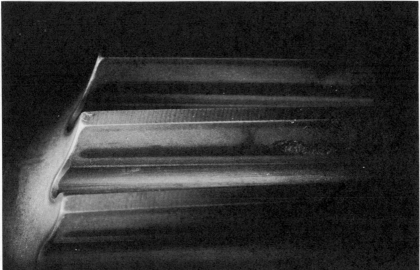

Figure 6.9 Surface temper indications.

detrimental to component strength or serviceability. Minor discontinuities in noncritical areas can be accepted without rework. When there is some question as to whether the discontinuity might propagate and cause distress, it can be removed by localized machining.

SURFACE TEMPER INSPECTION

When hardened components are ground there is the possibility of overheating the surface being machined and locally tempering the overheated areas. This may occur due to improper grinding practice or interruption of coolant flow during machining. If the surface temperature in a localized area exceeds the tempering temperature of the material, the area will soften. In the worst case the local temperature reaches a level sufficiently high to reharden the material, resulting in an area of brittle untempered martensite which can be a source of pitting during operation.

   To detect surface temper burns the component is dipped in an acid solution which etches the metal, exposing tempered areas. The process is usually known as Nital etch. Unburned areas of an etched part will be uniformly gray with a dull nonreflective surface. Tempering burns will appear as dark gray or black areas on the etched part. The darkness of the color reflects the severity of the burn. Rehardening burns will appear as white or light areas surrounded by a black tempered area.

   In order to set acceptance criteria for tempering, critical areas of the part being inspected must be defined. Also, examples of light temper distinguished by coloration must be available to use as comparisons. Critical gear areas are the loaded flanks of the gear teeth and the root fillet area. Clutch surfaces and bearing surfaces in rolling contact are also considered critical. No temper burns of any degree are allowed in the critical areas. Light temper covering less than 20% of a noncritical area may be acceptable. Rehardening burns are not acceptable in any area.

   Because temper is a surface discrepancy, burns can sometimes be removed by remachining the tempered area. In addition to exposing temper, the Nital etch will show up areas that are deficient in carbon or have had excessive stock removal during grinding. These areas will appear light in color when compared to a normal etched surface. Figure 6.9 illustrates tempered gear teeth; the darker color showing the tempered area, which in this case exhibited spalling.

REFERENCE

1. AGMA Standard 246.01A, Recommended Procedure for Carburized Aerospace Gearing, American Gear Manufacturers Association, Arlington, Va., November 1971.

7
MANUFACTURING METHODS

There are two basic methods of manufacturing gear teeth: the generating process and the forming process. When a gear tooth is generated, the workpiece and the cutting or grinding tool are in continuous mesh and the tooth form is generated by the tool. In other words, the work and the tool are conjugate to each other. Hobbing machines, shaper cutters, shaving machines, and many grinders use this principle.

When a gear tooth is formed, the tool is in the shape of the space that is being machined out. Some grinding machines use this principle with an indexing mechanism which allows the gear teeth to be formed tooth by tooth. Broaches are examples of form tools that machine all the gear teeth simultaneously. Gears of the type discussed in this book are initially cut on a hobbing machine or shaper cutter. The tooth forms are further refined by lapping, shaving, grinding, or honing. Following are descriptions of these various processes.

HOBBING

Figure 7.1 illustrates a hobbing machine and Figure 7.2 presents the hob itself. In this process the gear teeth are generated with the hob and workpiece rotating in a constant relationship while the hob is being fed into the work. Hobbing is a versatile and economical method of cutting gears. A hob of any given normal pitch and pressure angle will cut the teeth of all spur and helical gears having the same normal pitch and pressure angle. Hobs producing involute gears are basically straight sided and generate the involute form on the gear tooth by the meshing action.

Figure 7.1 Hobbing machine. (Courtesy of Barber Colman Corporation, Rockford, Ill.)

Figure 7.2 Gear hobs. (Courtesy of Barber Colman Corporation, Rockford, Ill.)

Figure 7.3 Shaper cutting machine. (Courtesy of Fellows Corporation, Springfield, Vt.)

Hobbing produces gears of qualities up to American Gear Manufacturers Association (AGMA) Quality Class 12. To achieve high accuracy the hobbing machine must be in excellent condition, the tooling holding the gear rigid and accurate, the gear blank accurate, and the hob of high quality. All types of materials can be hobbed. The great majority of hobbed gears are in a hardness range Rc 30 to 38; however, with special material cutters hardnesses up to Rc 60 can be machined. Hobbing will produce surface finishes in the range 16 to 64 μin. rms.

The only limitations of the hobbing process are the inability to machine internal gears and a requirement for axial space behind the gear teeth to allow the hob to run out. Gears that are adjacent to a shoulder or cluster gears cannot be hobbed.

Figure 7.4 Shaper cutters. (Courtesy of Fellows Corporation, Springfield, Vt.)

SHAPING

Figure 7.3 illustrates a shaper cutter machine and Figure 7.4 shaper cutters. Shaping is a generating process where the tool is in the form of a shape conjugate to the tooth being cut. When cutting involute gear teeth the shaper cutter is in the form of an involute gear which is hardened and has cutting clearance on the tooth sides. The gear blank and the cutter are rotated in the proper ratio while the cutter reciprocates axially through the gear blank. If a spur gear is being generated, the cutter will reciprocate through the workpiece in a straight path. To generate a helical gear the cutter must reciprocate in a helical motion which is imparted by a helical guide. This additional tooling required to cut helical gears is a disadvantage of shaper cutting compared to hobbing.

Shaping can be applied to internal as well as external spur or helical gears. Also, herringbone gears are cut by the shaping process. The shaper cutter does not require a large runout beyond the gear; therefore, it is a good method for cutting cluster gears or gears close to a shoulder. As with hobbing, the best shaper cutting can produce gears up to AGMA Quality Class 12 with finishes to 16 μin. rms. Hardnesses to Rc 43 can be shaper cut; however, the great majority of gears are cut in the range Rc 30 to 38.

LAPPING

Lapping is a method of correcting small errors in profile, lead, spacing, or runout of gear teeth. A gear can be lapped either by running it with its mating gear or meshing it with a lapping tool in the form of a gear. An abrasive lapping compound is introduced into the mesh to promote removal of metal. The abrasive medium must be uniformly spread across the teeth in an oil- or water-soluble base and should contain a rust inhibitor. Gear laps are usually made of cast iron into which the abrasive will embed.

For most lapped gears the process is short, typically minutes. Only small beneficial changes can be attained by lapping and too lengthy a cycle will usually destroy the profile. Because there is no sliding at the pitch line, lapping will tend to remove material above and below the pitch line only. An auxiliary axial sliding motion is incorporated in lapping machines to overcome this problem.

Wide face helical and double helical gears are often lapped for long periods of time such as several hours to improve their tooth contact. It has been found with wide face gears that have a helical overlap greater than 2 that no axial reciprocating motion is necessary while lapping.

Figure 7.5 Shaving cutter. (Courtesy of National Broach and Machine Division of Lear Siegler, Detroit.)

SHAVING

Shaving is a finishing operation which uses a high-speed-steel, hardened and ground, precision cutter which is in the form of a helical gear (Figure 7.5). The cutter teeth have gashes which act as cutting edges. Shaving will improve the tooth spacing, profile, lead, runout, and surface finish which was generated in the hobbing or shaping process. Shaving is often used to refine the tooth surface prior to hardening, thereby minimizing heat-treat distortion.

The shaving cutter is meshed with the work gear in a crossed axis relation-ship (Figure 7.6), and rotated while the center distance between the two is reduced in small increments. Simultaneously, the work is traversing back and forth in relation to the cutter.

Shaving can achieve gear qualities of up to AGMA Class 13. The quality of a shaved gear is strongly dependent on the quality of the preceding hobbing or shaping operation. Shaving can remove approximately 70% of the errors in the as-cut gear. A surface finish of 25 μin. rms is normally achieved, with much finer finishes possible. The optimum gear tooth hardness for the shaving process is Rc 30.

Figure 7.6 Shaving cutter machine. (Courtesy of National Broach and Machine Division of Lear Siegler, Detroit.)

Shaving may be applied to spur and helical gears, both external and internal. Tooth profile modifications and crowning can be achieved. Profile modifications are accomplished by grinding the correct form into the cutting tool teeth. Crowning is accomplished by sinking the cutter more deeply into the tooth ends than into the middle.

GRINDING

The most accurate gears are produced by the grinding process. It is used primarily to finish hardened gears in the range Rc 55 and up. Grinding can produce AGMA Qualities of Class 14 and higher and surface finishes as fine as 10 μin. rms. Gear tooth profile modification and crowning of the gear tooth face can be accurately produced by the grinding process. The two basic techniques used to grind gear teeth are the form grinding method and the generating method. Either method, when closely controlled, can produce the highest-quality gears.

Figure 7.7 illustrates a form grinding machine which utilizes a disk-type grinding wheel. The wheel form is contoured by a diamond dressing tool into the

Figure 7.7 Form grinding machine. (Courtesy of National Broach and Machine Division of Lear Siegler, Detroit.)

form of the tooth space being machined out. Both the root and the tooth flank area can be ground, or the flank area alone, by cropping the tip of the grinding wheel. The wheel grinds one tooth at a time and an index plate having the same number of spaces as the number of teeth in the gear is used to index around the part. Tooth spacing is a function of the accuracy of the index plate and spacing error can be minimized by using an index plate larger in diameter than the part being ground.

Generating grinders use disk wheels of various types, such as conical or saucer shaped or threaded worm types of wheels. The work is rolled with respect to the grinding wheel and the wheel reciprocates axially with respect to the work. Generating grinder wheels act as though they were straight-sided racks in mesh with the gear being ground. The disk is therefore dressed in a straight-sided form with the proper pressure angle. The straight-sided form is modified to achieve whatever tooth modifications are required on the gear to be ground.

When grinding case-hardened gears it is important to control all phases of the process in order to minimize the amount of stock removal during grinding. Too much stock removal will result in loss of the hardened surface area and the benefits of case hardening. This is particularly critical in the root fillet area. In order to have sufficient case depth in this critical bending stress region, the tooth must be cut prior to hardening with a generous root fillet radius. Figure 7.8A shows the shallow case depth that results in the root fillet area when the radius is too small. The surface area available for carbon to penetrate in this region is insufficient. After hardening, when the tooth is ground, little or no case will be left and bending failure resistance will be significantly reduced. In order to attain the largest possible root fillet radius the hob or shaper cutter tips must be designed to produce an undercut below the form diameter of the gear tooth (Figure 7.8B). This configuration is referred to as protuberance cut

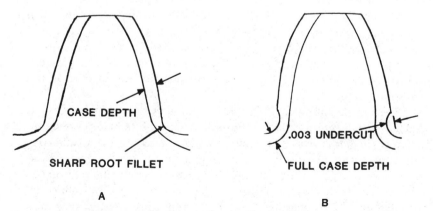

Figure 7.8 Hardened gear teeth in the as-cut condition.

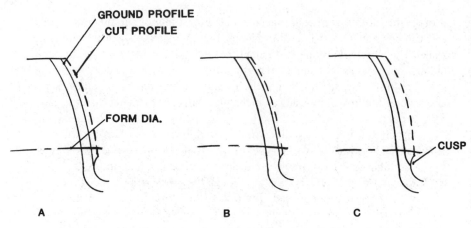

Figure 7.9   Hardened gear teeth after grinding.

since in order to generate the undercut the cutter has a protuberance added at its tip. The case depth in the root fillet is greater with protuberance cut teeth because there is a greater surface area available for carbon to penetrate during the carburizing cycle. Another advantage of protuberance cutting is that only the tooth flank need be ground, leaving the carburized case in the root fillet intact. Figure 7.9 shows the types of tooth profiles that result from this grinding technique. Assume that the as-cut undercut is 0.003 in., as shown in Figure 7.8B. If the stock removal during grinding of the tooth flank is 0.003 in., the profile will be tangent to the root fillet radius (Figure 7.9A). This is the ideal case. Stock removal of less than 0.003 in. will leave some undercut in the finished tooth (Figure 7.9B) and stock removal greater than 0.003 in. will result in a cusp (Figure 7.9C).

There is some question whether a tooth with unground root or one with a fully ground root is stronger. The fully ground root may have a better form, however, some of the hardened case has been removed. With a fully ground root there is always the danger that stock removal has been excessive. Both grinding methods are commonly used and, if correctly done, produce gears that achieve successful operation.

The grinding operation is strongly influenced by the condition of the gear after heat treatment. In carburized parts excessive case carbon can result in temper or grinding cracks during grinding. Oxidation resulting in decarbuziation at the surface will cause overheating during grinding and surface temper. Heat-treat distortion must be minimized so that excessive stock removal is not required to clean up the part. Often, development programs are required to determine what type of distortion will occur during heat treating. The parts can then be cut to take advantage of the movement during heat treating. Use of quenching dies is another method to limit heat-treat distortion.

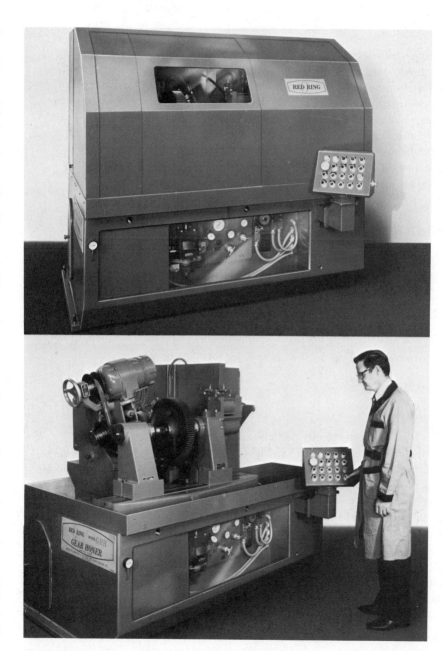

Figure 7.10 Honing machine. (Courtesy of National Broach and Machine Division of Lear Siegler, Detroit.)

HONING

Honing is a finishing process for hardened gears in which an abrasive impregnated plastic helical gear-shaped tool meshes with the work gear in a crossed axis relationship (Figure 7.10). As the tool and work rotate the honing tool traverses across the workpiece. Honing is used to improve gear tooth surface finish, remove nicks and burrs on the teeth, and correct minor errors in the gear tooth shape. Honing can produce surface finishes as fine as 6 μin. rms. It can be applied to external and internal gear teeth and achieve profile modification and crowning. The amount of stock removed in the honing process is small: 0.0003 to 0.001 in.

OTHER METHODS OF GEAR TOOTH MACHINING

The techniques used to generate and form gear teeth discussed above are not the only machining processes in use. Gears are also produced by milling, shear cutting, broaching, stamping, molding, die casting, sintering, and rolling. These processes either do not produce gears of sufficient strength or accuracy for the applications covered in this book or are high-production techniques which require considerable investment in tooling and are not economical for the relatively small production lots run for this type of gearing.

SHOT PEENING

In this process high-velocity particles are shot at the gear tooth, deforming the metal surface, in order to produce residual compressive stresses at the surface. The depressions are very shallow: less than 0.0001 in. Shot peening is used primarily to induce or increase the compressive stress in the root fillet area to a level sufficiently high such that the bending stress applied by the tooth load is offset and fatigue cracks at the surface will not initiate or propagate. Similar beneficial results in pitting resistance are claimed for shot peening of the tooth flanks; however, some users maintain that shot peening increases tooth surface roughness and they shot peen the root fillet area only. This may be accomplished by masking the gear tooth flank during peening or by shaving or grinding the tooth flanks after peening.

Effectiveness of shot peening depends on the following factors:

Type of shot
Hardness of shot
Uniformity of shot

Velocity of the shot stream
Duration of the treatment
Distance from the nozzle to the workpiece
Angle of impact

In order to develop the right combination of these factors for a given application a technique named the Almen strip test is used. The test strip is a standard piece of spring steel which is shot peened on one side. The resulting residual surface compressive stresses make the strip bow upward and the height of the bowed arc is an index of the intensity of the peening. The arc height can be related to the depth of the residual stress layer and the magnitude of residual stress.

Steel shot with a hardness of Rc 45 to 55 is commonly used for through-hardened gears and a shot hardness of Rc 55 to 65 is used for case-hardened parts. In some cases glass shot is used. In order to penetrate the root fillet area the shot diameter should be no larger than half of the smallest fillet radius. The shot quality in terms of spherical shape and uniform size is important. Sharp-edged particles can damage the tooth surface. The shot stream should be approximately perpendicular to the surface being blasted. Because the gear tooth root is contoured, it is usually necessary to blast from more than one position.

INSPECTION EQUIPMENT

The gear tooth involute profile, lead, spacing, runout, and thickness can be accurately measured on the bench using standard inspection equipment; however, specialized machines are available to perform these inspections quickly, economically, and with the capability of recording the results on a permanent chart. When all the foregoing elements of a gear tooth are measured individually, the process is called an analytical inspection. Another technique used is the functional inspection, where the gear in question is rolled with a master gear or its mating gear to determine acceptability. When the functional inspection indicates a problem, an analytical inspection is performed to determine the cause.

The following paragraphs will describe some of the specialized instruments available for performing analytical and functional inspections of gear teeth.

Involute Profile Measuring Machine

Figure 7.11 illustrates one type of involute checker. The gear is mounted in centers and rotates in a timed relationship with a stylus that traverses the

Figure 7.11 Involute profile measuring machine. (Courtesy of Fellows Corporation, Springfield, Vt.)

profile in the transverse plane. The deviation of the profile from a true involute form is measured and displayed on a dial indicator graduated in 0.0001 in. increments and/or a recording pickup which may have a sensitivity as fine as 0.000020 in. The profile is recorded on a strip chart. To calibrate the machine an accurate master with known form is inspected.

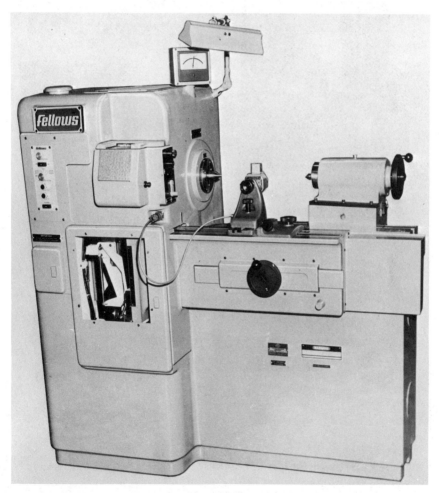

Figure 7.12 Lead checking machine. (Courtesy of Fellows Corporation, Springfield, Vt.)

Figure 7.13 True-position spacing checker. (Courtesy of Fellows Corporation, Springfield, Vt.)

Lead Measuring Machine

A gear tooth lead inspection is an indication of the alignment of the face with the axis of rotation. In the case of a spur gear the face width should be parallel to the reference axis. For helical gear teeth the lead check determines if the tooth has the required helix angle. Figure 7.12 illustrates a lead checking machine. The gear is mounted between centers and a stylus traverses the face in an axial direction. For helical gears the gear rotates as the stylus is advanced at the proper rate to satisfy the lead relationship. If the lead is correct the stylus will describe a straight line. The deviation from a straight line is displayed on a dial indicator and/or a strip chart using an electrical pickup. The lead measurement can be made to an accuracy of 0.0001 in. or better.

Tooth Spacing

Two types of tooth spacing checkers are in general use. One measures pitch variation, the difference between one circular pitch, and the circular pitches immediately before and after. This type of machine has a fixed finger that acts as a stop on the pitch line of a tooth. A second finger or stylus senses the position of the adjacent tooth and actuates a dial indicator or recording pen.

The other type of tooth spacing checker measures index variation or the true position of a tooth. The gear is mounted on a rotating disk which turns in increments corresponding to the number of teeth in the gear. A stylus comes in and contacts each tooth near the pitch line. The first tooth is set at 0.0 and the deviation of each subsequent tooth from its theoretical position is measured and recorded. Figure 7.13 illustrates a true-position spacing checker. The two types of spacing checkers will not give identical readings on the same gear because they operate on different principles; however, analysis of the findings of either machine will lead to the same conclusions concerning gear acceptability. On either machine tooth spacing can be checked to accuracies of 0.0001 in. or better.

FUNCTIONAL INSPECTIONS

Composite Inspection

Figure 7.14 illustrates a composite gear checking machine. This type of inspection is sometimes called a red line, in reference to the strip chart that is produced. The gear to be checked is mated with a master gear which is mounted on springs such that the center distance between the two gears can vary as the teeth mesh. The variation in center distance is a function of tooth error and is measured and recorded. The variation in center distance from tooth to tooth is a composite measurement of errors in tooth spacing, profile, lead, and surface

Figure 7.14 Composite gear checking machine. (Courtesy of Fellows Corporation, Springfield, Vt.)

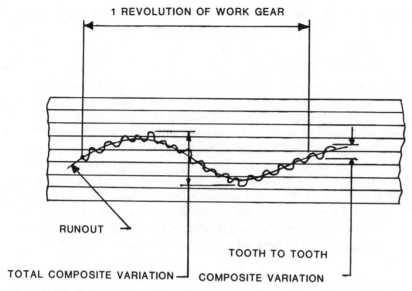

Figure 7.15 Typical red line chart.

finish. The total composite variation is a measure of the gear runout plus the local composite variations. Figure 7.15 illustrates a red line chart.

The composite error reading is influenced by errors in the master gear as well as errors in the part being inspected; therefore, the master must be as perfect as possible. The composite inspection method is a quick, inexpensive way of checking production gears. Analysis of the red line chart can differentiate between the various types of errors; however, the composite check is not sufficiently accurate for measurement of critical gears above AGMA Quality Class 12 or gears that require accurate profile or lead modifications.

Error in Action

Gear tooth inaccuracies will result in variations of velocity ratio in a pair of mating gears as the driven gear accelerates and decelerates. Machines are becoming available which measure this variation. One type uses friction disks with diameters machined to the exact pitch diameters of the gears to be inspected. One disk is attached to the driving gear and the other disk is mounted on the same axis but independent of the driven gear. The driven disk operates at a constant speed and the driven gear angular velocity is compared to it electrically and the variation recorded. This type of checker measures the action of the loaded faces of the gear teeth only, unlike the composite checker, where the mating gears are always in tight mesh with both faces touching. For this reason it is sometimes referred to as a single flank tooth tester.

Tooth Pattern Inspection

Often gear teeth are rolled together after a marking compound is applied to the teeth in order to check the contact pattern. This can be done in an inspection fixture or in the actual gear housing. In an inspection fixture the tooth pattern will show the amount of contact across the face width and indicate how well the teeth line up. In the gear housing errors other than tooth alignment come into play, such as bearing bore parallelism. Sometimes the pattern check is used to determine a further modification of the gear tooth, which is then remachined to achieve good contact. When performing this type of check care must be taken to spread the marking compound very lightly. If the thickness of marking compound is greater than approximately 0.0001 in., the errors that the inspection is attempting to reveal will be masked by excessive smearing of the compound.

BIBLIOGRAPHY

Dudley, D. W., *Gear Handbook*, McGraw-Hill, New York, 1962, Chaps. 16 to 23.
Modern Methods of Gear Manufacture, National Broach and Machine Division, Lear Siegler, Inc., Detroit, 1972.

8
GEARBOX ECONOMICS

There are a multitude of variables affecting gearbox costs which make it difficult to determine how expensive a unit must be for a given application. The trade-off between design conservatism and cost is always present, the goal being to achieve satisfactory performance at minimum price. In each application different criteria are applied as to what constitutes satisfactory performance. For a relatively inexpensive gearbox some period of trouble-free performance after which the unit is easily replaced may be perfectly satisfactory. In some process industry applications the cost of downtime incurred if a gearbox must be modified or replaced is so great that the initial gearbox price in comparison is insignificant. In such a case satisfactory performance is equated with extremely high reliability over long periods of time and in order to achieve this reliability, unit costs increase significantly.

The cost of a gearbox can be divided into three elements:

Material costs
Manufacturing costs
Purchased items such as bearings, seals, lubrication components, and so on.

These three factors are dependent on the gearbox design and configuration chosen. To illustrate this, let us look at an example of a 500-hp gear unit driven by a 5000-rpm steam turbine and driving an 1800-rpm generator. This application might be satisfied by any of the following designs:

1. Through-hardened cut double helical gearing
2. Through-hardened cut and shaved double helical gearing
3. Single helical, case hardened and ground gearing

4. Single helical, case hardened and ground pinion meshing with a through-
    hardened cut gear

Design 1 will be the largest unit since the tooth stresses must be low for this relatively inaccurate method of fabrication. Design 3 will be the smallest and therefore will have significantly lower material costs. The single helical gearing,
however, requires more machining time since it is cut and ground and also
requires more sophisticated heat treatment. Bearings for design 3 will probably
be more expensive since there are gear thrust loads to deal with. The larger gear
sets of designs 1 and 2 will generate more heat in churning, requiring more
cooling oil flow and therefore a larger pump, cooler, filter, reservoir. These are
only a few examples of many tradeoffs that are made in selecting a gearbox
design.

   If one were to solicit competitive quotes from companies making each of
these types of designs it would not be possible to predict which would be least
expensive. If one company's design were inherently significantly more costly
than the others, the company would not long remain in business. Quite often the
low bid is not a reflection of economical design or manufacture but an indication of how badly the company wants the particular program.

   A cost consideration beyond the gearbox itself is the impact of the unit on
the total system. A smaller gearbox will save costs associated with the base plate.
Also, shipping costs will be less. The gearbox configuration, parallel shaft with
an offset between input and output, or planetary with concentric shafts, will
also affect the system design and influence costs. Planetary gears lend themselves
to close coupling of the gearbox to the driver or driven equipment, with the
potential of saving coupling and shafting costs.

   From the discussion above it can be seen that if one wanted to determine
the most cost effective gearbox for a new application and were free to choose
any manufacturing and heat treating methods, a detailed design study and cost
analysis of the numerous potential design solutions would be required to arrive
at a sensible conclusion. More often companies buy gearboxes complete from a
manufacturer with existing facilities or have their own machine tools and
facilities; therefore, the design chosen reflects the type of gearbox the company
has made in the past. The cost question then focuses on what materials to use
and what degree of quality to require in the gearbox components.

GEAR QUALITY

The types of gears being discussed in this book range in American Gear Manufacturers Association (AGMA) Quality Class from about 8 for the less accurate end
to 14 for the most accurate. A Quality Class 8 gear might be used in low-speed
mining equipment, while a Quality Class 14 gear might be used in a sophisticated

very high speed compressor drive. Automotive gears are typically Quality Class 10 to 12. The quality achieved is to some extent determined by the finishing process; however, the various processes overlap in terms of quality produced. An approximate comparison follows:

AGMA quality range	Process
8-12	Hobbing and shaping
9-13	Shaving
10-14 and up	Grinding

Figure 8.1 presents an approximate relationship between AGMA quality class and cost. The numbers are relative, with a Quality 14 gear costing approximately 8 times as much as a Quality Class 8 gear. The figure shows that at the high-quality end of the spectrum a small increase in quality requires a large expenditure. Conversely, at the low end decreased quality offers little savings. AGMA standards give recommendations for quality classes suitable for various applications. As experience is gained in a particular service it is possible to ascertain more accurately which tooth tolerances are important and how closely they must be held. It would then be appropriate to buy gears to a specification reflecting that experience rather than to a general quality specification. To be

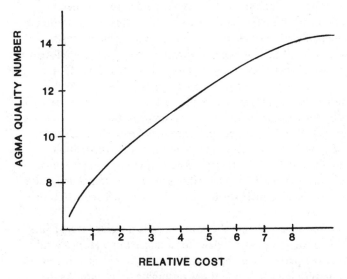

Figure 8.1 Approximate relationship between AGMA quality class and cost.

certain that the quality specified is reflected in the gear tooth geometry, a user should request inspection documentation. These might take the form of red line, involute, lead, or tooth spacing charts. Measurements over balls, bore parallelism, center distance measurements, and so on, might also be requested to be documented. There may be additional costs for this documentation. The need for and extent of inspection documentation is a matter of judgment, but if AGMA quality class requirements of 12 or above are specified, it is recommended that 100% documentation be requested. For Quality Class 14 the user and manufacturer should agree beforehand on how the inspections will be made and the results interpreted. For quality classes lower than 12 it is good practice to document prototype units so that if field problems occur in subsequent production gearboxes, a baseline of quality has been established.

So far, our discussion of quality has been confined to gear tooth geometry. If it is determined that high quality is required in the gear teeth, undoubtedly other components in the unit must also be specified to equivalent quality levels. In fact, failures associated with bearings, the dynamic system, and the lubrication system are more common than gear tooth problems; therefore, complete specification of all components in the unit in addition to the gears is necessary.

MATERIAL COSTS

Gear steels are bought on a dollars per pound basis either as bar stock or in a forged state. Up to a diameter of approximately 6 in. it is cheaper to buy bar stock than forgings. Above 6 in. bar stock becomes harder to procure and forgings have an economic advantage, particularly since forgings can be worked closer to the final shape than bar stock and therefore save machining costs.

Taking a low-alloy steel such as Society of Automotive Engineers (SAE) 1040 as a base, the cost of high-alloy steels used in gears such as SAE 4340 or 9310 may be three or four times as high. Of course, gears made of high-alloy steel will usually be smaller than those of low-alloy steel, so the cost of material is somewhat offset by the smaller quantity. If a requirement for vacuum-melt steel rather than air-melt steel is specified, the price may be doubled. Over the years the quality of gear steels has deteriorated and it is not uncommon to encounter impurities or voids in air-melt steels. These discrepancies are sometimes found only after several machining operations have been completed; therefore, the higher cost of vacuum-melt steels may be offset by a reduction of scrapped work in process.

The question often arises: Should the gear casing be a welded steel fabrication or a casting? From a cost point of view, for a single unit a fabrication is cheaper. This is because a pattern must be machined prior to pouring the casting, and developing the pattern is a costly and time-consuming procedure. As the quantity of gearboxes required increases, pattern costs can be amortized over the production run and castings become economically attractive. The casting can be poured close to the final shape, and extensive machining and welding can be eliminated. For a typical gear unit, the break-even point may be as low as three gearboxes. Castings may be made of cast iron or steel. A steel casting costs approximately 30% more than a cast iron part. On occasion, gear housings are made of aluminum to save weight. Aluminum is twice as expensive as steel on a dollar per pound basis, but because the housing weight will be much lower, it is not clear which is more expensive. Aluminum housings, however, usually have to incorporate steel bearing liners and inserts for threads, which increase cost.

EFFECT OF QUANTITY ON COSTS

When manufacturing gearbox components, the time to set up the equipment far exceeds the time required to machine the parts. Therefore, it is far more expensive to perform a machining operation on one part than on many parts. For instance, it may take 8 hr to set up a gear grinder and 2 hr to grind a part. Assuming a cost of $X/hr, one part would cost $10(X), whereas four parts would cost $4(X) each $[(8/4 + 2)(X)]$.

If a single special unit is required for an application, design and tooling costs must be considered. These costs will probably exceed the manufacturing expenses. It is therefore worthwhile to attempt to use a standard gear unit or one that has been manufactured for a different application, even if it requires compromising the system design to some extent.

BIBLIOGRAPHY

Hamilton, J. M., Are You Paying Too Much for Gears, *Machine Design*, October 1972, pp. 144–150.

Kron, H. O., Optimum Design of Parallel Shaft Gearing, ASME Paper 72-PTG-17, October 1972.

The Cost of Gear Accuracy, *Design Engineering*, January 1981, pp. 49–52.

AGMA Gear Handbook 390.01, Vol. 1, Gear Classification, Materials and Measuring Methods for Unassembled Gears, American Gear Manufacturers Association, Arlington, Va., 1973.

9

PLANETARY GEAR TRAINS

A planetary gear train is one in which the power is transmitted through two or more load paths rather than the single load path of a simple gear mesh. Figure 9.1 illustrates the components of the simplest type planetary, a sun gear, planet gears, a ring gear, and the planet carrier. When the carrier rotates about the center of the system a point on a planet gear not only rotates about the axis of the planet gear but also about the center of the system and this type of drive is called epicyclic.

In the United States, planetary gears are widely utilized in automotive automatic transmissions and in aerospace drives such as turbine engine reduction gears or helicopter transmissions but are not used extensively in industrial applications. There is a trend toward increasing utilization of planetary gears because they offer the following advantages:

1. Because they share load between several meshes, planetaries are more compact than parallel shaft drives and offer significant envelope and weight savings. An example illustrates this point:

> Let us take a 1000-hp electric motor at 1800 rpm driving a compressor at 7200 rpm. The required American Gear Manufacturers Association (AGMA) service factor is 1.6.
> A hardened and ground, helical, parallel shaft gear set would have the following dimensions:

> Pinion pitch diameter: 4.0 in.
> Gear pitch diameter: 16.0 in.
> Face width: 4.0
> Gearbox envelope: 24 X 12 X 12 in.
> Gearbox weight: 650 lb

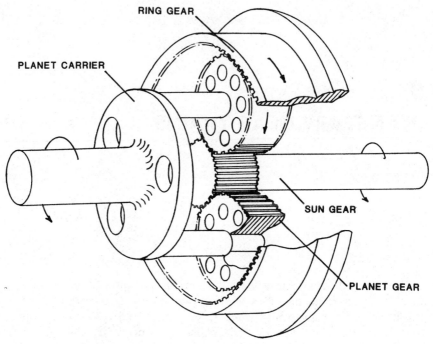

Figure 9.1 Simple planetary gearset components.

The equivalent planetary gear set with a stationary ring gear and rotating carrier would have the following dimensions:

Sun pitch diameter: 3.5 in.
Ring pitch diameter: 10.5 in.
Planet pitch diameter: 3.5 in.
Face width: 3.0 in.
Gearbox envelope: 15 in. diameter × 10 in. length
Gearbox weight: 250 lb.

Figure 9.2 illustrates the gearbox size comparison.

2. In addition to achieving minimum weight and envelope, the relatively smaller and stiffer components which result from the use of planetary gearing lead to reduced noise and vibration and increased efficiency. Noise and efficiency are strongly dependent on the speed of the components. In the example above, the parallel shaft pitch line velocity is 7540 fpm compared to the planetary gear pitch line velocity of 4948 fpm, a significant difference. The lower planetary pitch line velocity is a result of the smaller sun gear and the epicyclic

PARALLEL SHAFT CONFIGURATION   PLANETARY CONFIGURATION

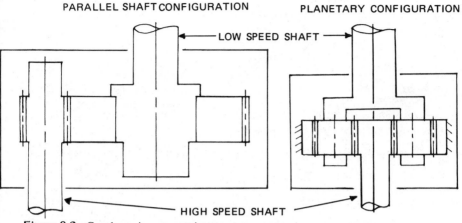

Figure 9.2 Gearbox size comparison.

action of the gear mesh. Later in this chapter the method of calculating epicyclic speeds is derived.

3. The input and output shaft axes in planetary gearing are concentric. This can lead to space savings in some installations by allowing the driving and driven equipment to be in-line. The coaxial feature is very important in automatic transmissions since it makes rapid speed changes possible without the necessity of taking gears out of mesh. The rotating components are controlled by the use of clutches and brakes to achieve speed changes.

4. The resultant radial forces on the input and output shafts of planetary gearboxes are zero since the arrangement cancels out all radial forces. In effect, the planetary gearbox transmits only torque. This simplifies bearing design and in some applications allows close coupling of the gearbox to the driving or driven equipment.

ANALYSIS OF PLANETARY GEAR ARRANGEMENTS

Planetary gears can be arranged in a multitude of configurations to achieve specific ratios and power splits. This section presents a method of calculating planetary speed ratios and power flows. The method is applied to several examples and by following the logic the reader can generate equations describing the mechanics of any planetary configuration. The following nomenclature will be used throughout.

Symbols
  R = gear pitch radius, in.
  W = gear angular velocity, rad/sec

W_T = tangential load, lb

n   = gear rpm

Subscripts

s   = sun gear

p   = planet gear

r   = ring gear

c   = carrier

1   = primary stage (high-speed stage)

2   = secondary stage (low-speed stage)

For example, the symbol R_{s_1} would be the pitch radius of the primary-stage sun gear.

Simple Single-Stage Planetary

Figure 9.1 shows the general case where the sun gear, ring gear, and planet carrier are all free to rotate. Directions of rotation are assumed to be as shown in the figure. With all members rotating this is termed a differential system and there are potentially six different ways to connect prime movers and driven equipment:

Inputs	Outputs
Sun	Carrier, ring
Carrier	Sun, ring
Ring	Carrier, sun
Sun, ring	Carrier
Carrier, sun	Ring
Ring, carrier	Sun

The rotating elements in each of the arrangements above have a distinct speed and torque relationship. In order to define the angular velocity of all three elements in the planetary gear train, the angular velocity of two elements must be specified. Let us derive the generalized speed relationships of the gear train in Figure 9.1.

1. A point on the pitch diameter of the sun gear has a tangential velocity of $W_s R_s$.
2. The point on the sun gear pitch diameter is meshing with a point on the planet gear pitch diameter which has a tangential velocity made up of two components, $W_p R_p + W_c R_s$.

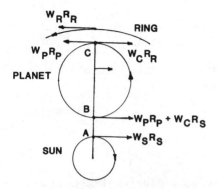

Figure 9.3 Idealized planetary system.

Figure 9.3 shows the planetary system idealized as friction disks with a point A on the sun disk meshing with a point B on the planet disk and also shows their respective tangential velocities. Because the points are meshing they must have the same tangential velocities; therefore,

$$W_s R_s = W_p R_p + W_c R_s \tag{9.1}$$

3. At the ring gear mesh a similar equation may be written:

$$W_r R_r = W_p R_p - W_c R_r \tag{9.2}$$

Note that at point C on the planet gear (Figure 9.3) the components of the tangential velocity are in opposite directions and subtractive, whereas at point B they are in the same direction and additive. The tangential velocity of the planet carrier is zero at the center of the system and equals $W_c R$ at any radius R. Combining Eqs. (9.1) and (9.2), the general speed ratio equation is

$$W_s R_s = W_r R_r + W_c (R_s + R_r) \tag{9.3}$$

Let us apply this equation to the example shown earlier in this chapter:

$$R_s = \frac{3.5}{2}$$

$$R_p = \frac{3.5}{2}$$

$$R_r = \frac{10.5}{2}$$

In this case the ring gear was stationary; therefore,

$$W_r = 0$$

$$\text{Ratio} = \frac{W_s}{W_c} = \frac{R_s + R_r}{R_s} = 4$$

In a simple gear train the rpm of the planet gear would be rpm sun $\cdot \left(\dfrac{R_s}{R_p}\right)$;
however, because of the epicyclic action the planet rpm is reduced and can be calculated from either Eq. (9.1) or (9.2) with

rpm sun   = 7200

and

rpm carrier = 1800

the planet rpm = 5400 and the pitch line velocity of the sun planet mesh is

$$\text{PLV} = \frac{5400(3.5)\pi}{12} = 4948 \text{ fpm}$$

The rolling velocity of the sun with the planet is 5400(3.5/3.5) = 5400 rpm and its absolute velocity is 7200 rpm.

If, in the system shown in Figure 9.1, the carrier is held stationary, $W_c = 0$ and Eq. (9.3) becomes

$$\frac{W_s}{W_r} = \frac{R_r}{R_s}$$

In this case, where there is no epicyclic action, the speed ratio is simply the ratio of the sun and ring gear radii. This configuration is sometimes called a star system.

If the sun gear is held stationary, $W_s = 0$ and Eq. (9.3) becomes

$$\frac{W_r}{W_c} = -\frac{R_r + R_s}{R_r}$$

The negative sign indicates that the sense of rotation of one of the components shown in Figure 9.1 was chosen opposite to the direction in which it actually rotates. In this case, if the ring gear is driving, the carrier would rotate in the same direction as the ring gear rather than the direction shown in Figure 9.1 (viewed from the sun gear end).

In the equations above rpm can be used in place of angular velocity and number of teeth in place of pitch radius. Quite often, planetary gear equations are presented in this manner.

If all components of a planetary gear set are free to rotate, the speed ratios are dependent on the power split between the components. Figure 9.4 shows the tangential tooth loads at the sun and ring gear mesh points. The torque and horsepower of each component are as follows:

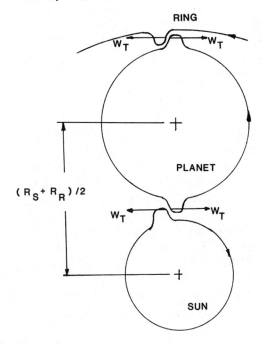

Figure 9.4 Planetary tangential tooth loads.

Component	Torque (in.-lb)	Horsepower
Sun	$W_T R_s$	$\dfrac{W_T R_s n_s}{63,025}$
Carrier	$2W_T \dfrac{R_s + R_r}{2}$	$\dfrac{W_T (R_s + R_r) n_c}{63,025}$
Ring	$W_T R_r$	$\dfrac{W_T R_r n_r}{63,025}$

It should be noted that the tangential load W_T used in the table is the summation of the tangential loads at each planet gear mesh.

Let us assume that in the previous example the 1800 rpm carrier with a 1000 hp input drives a compressor which is attached to the sun gear and also drives an oil pump which is attached to the ring gear. The oil pump absorbs

50 hp. The three horsepower equations in the previous table can be solved for the three unknowns W_T, N_s, and R_r. Using the carrier horsepower equation with R_s = 1.75, R_r = 5.25, and n_c = 1800 the tangential load W_T = 5002 lb. Knowing W_T, the sun and ring horsepower equations yield

$$n_s = 6840 \text{ rpm}$$
$$n_r = 120 \text{ rpm}$$

It should be noted that in this discussion no attempt was made to account for the power loss due to friction in the train.

The arrangement shown in Figure 9.1 is capable of producing gear ratios in a range of approximately 2:1 to 10:1. The speed ratio derived earlier in this section is R_r/R_s for the stationary carrier, rotating ring gear case. In this situation the input and output directions of rotation are reversed. With a stationary ring gear and rotating carrier, the speed ratio is $1 + (R_r/R_s)$ and the input and output rotation directions are the same. At low ratios the sun gear diameter approaches the ring gear diameter and therefore planet gear diameters become small and impractical. As one strives for the higher ratios in a single-stage planetary, the sun gear becomes smaller and is stress limited. Also, at ratios above approximately 10 the ring gear becomes large and more economical arrangements are available.

Compound Planetary

For a ratio of approximately 10:1 to 16:1, the compound planetary gear set (Figure 9.5) provides a practical and economic configuration. The speed reduction equations for this design follow:

$$W_s R_s = W_{p_1} R_{p_1} + W_c R_s$$
$$W_r R_r = W_{p_2} R_{p_2} - W_c R_r$$
$$W_{p_1} = W_{p_2}$$

Figure 9.5 Compound planetary gear.

Combining these equations, we have

$$W_s = W_r \frac{R_r R_{p1}}{R_s R_{p2}} + W_c \left(1 + \frac{R_r R_{p1}}{R_s R_{p2}}\right)$$

In the case of a stationary ring gear and rotating carrier the input and output direction of rotation are the same and the speed ratio is

$$R = \frac{W_s}{W_c} = 1 + \frac{R_r R_{p1}}{R_s R_{p2}}$$

In the case of a stationary carrier and rotating ring gear the input and output directions of rotation are opposite and the speed ratio is

$$R = \frac{W_s}{W_r} = \frac{R_r R_{p1}}{R_s R_{p2}}$$

For low ratios in the range 3:1 a reverted type of compound planetary is used as shown in Figure 9.6. The speed ratio for this train is simply

$$R = \frac{W_{s1}}{W_{s2}} = \frac{R_{s2} R_{p1}}{R_{s1} R_{p2}}$$

In this case the input and output directions of rotation are reversed.

Multistage Planetaries

By combining stages of planetary gearing in various arrangements, large ratios can be achieved. For instance, let us take two stages of simple planetary gearing as shown in Figure 9.1 and combine them in two ways as shown in Figure 9.7A and 9.7B. Figure 9.7A presents a drive where the primary-stage carrier drives the secondary-stage sun gear. The reduction ratio of the primary stage is

$$R_1 = \frac{W_{s1}}{W_{c1}} = \frac{R_{s1} + R_{r1}}{R_{s1}}$$

The reduction ratio of the second stage with a carrier output is

$$R_2 = \frac{W_{s2}}{W_{c2}} = \frac{R_{s2} + R_{r2}}{R_{s2}}$$

and the total reduction is

$$R = R_1 R_2 = \left(1 + \frac{R_{r1}}{R_{s1}}\right)\left(1 + \frac{R_{r2}}{R_{s2}}\right)$$

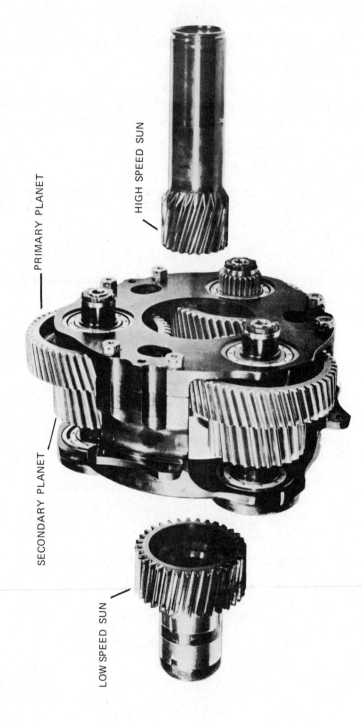

PRIMARY PLANET

HIGH SPEED SUN

SECONDARY PLANET

LOW SPEED SUN

Figure 9.6 Reverted compound planetary. (Courtesy of AVCO, Lycoming Division, Stratford, Conn.)

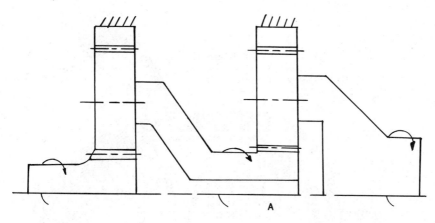

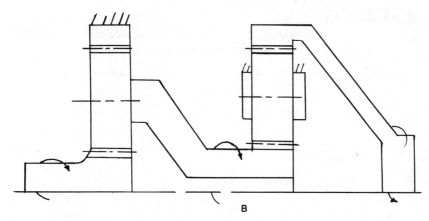

Figure 9.7 Two-stage planetary gearing.

The input and output rotation are in the same direction. Figure 9.7B shows a ring gear output from the second stage and in this case the second-stage reduction ratio is

$$R_2 = \frac{W_{s2}}{W_{r2}} = \frac{R_{r2}}{R_{s2}}$$

and the total reduction ratio is

$$R = R_1 R_2 = \left(1 + \frac{R_{r1}}{R_{s1}}\right) \frac{R_{r2}}{R_{s2}}$$

The input and output directions of rotation are opposite in sense.

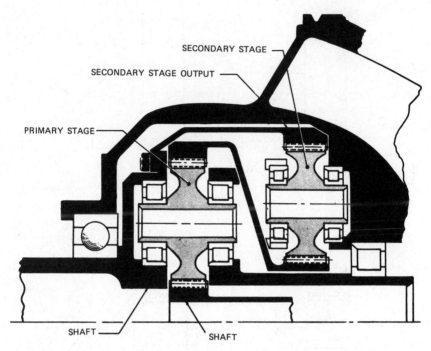

Figure 9.8 Split power transmission.

Split Power Transmissions

An interesting planetary arrangement which results in a very compact design is the split power transmission. Figure 9.8 illustrates one version which is most useful in a ratio range of approximately 15:1. As shown in the figure the primary planetary stage is driven by the sun gear and has both a rotating carrier and ring gear. The primary ring gear drives the sun gear of the second stage which through the stationary second-stage planets drives the second-stage ring gear. Both the second-stage ring gear and the primary stage carrier are connected to the output shaft, thus enabling the power split. Part of the input power is transmitted directly to the output shaft through the primary stage carrier, thereby bypassing the second stage of gearing. Following are the speed and power equations describing the split power transmission:

$$W_{s_1} R_{s_1} = W_{p_1} R_{p_1} + W_{c_1} R_{s_1}$$

$$W_{r_1} R_{r_1} = W_{p_1} R_{r_1} - W_{c_1} R_{r_1}$$

$$W_{r_1} = W_{s_2}$$

$$W_{s2} R_{s2} = W_{p2} R_{p2} = W_{r2} R_{r2}$$

$$W_{r2} = W_{c1}$$

where the subscripts $s_1, c_1, p_1,$ and r_1 refer to the primary-stage sun, carrier, planet, and ring and the subscripts $s_2, p_2,$ and r_2 refer to the second-stage sun, planet, and ring gears.

Combining the equations above we arrive at the expression for the speed ratio:

$$R = \frac{W_{s2}}{W_{c1}} = \frac{W_{s1}}{W_{r2}} = 1 + \left(\frac{R_{r1}}{R_{s1}}\right) + \left(\frac{R_{r2}}{R_{s2}} \frac{R_{r1}}{R_{s1}}\right)$$

To determine the horsepower split between the primary and secondary stage, the following expressions are derived:

$$HP_{in} = \frac{W_T R_{s1} n_{s1}}{63,025}$$

where n_{s1} is the primary sun gear rpm.

$$HP \text{ primary carrier} = 2 W_T \frac{(R_{s1} + R_{r1}) n_c}{2(63,025)}$$

where n_c is the carrier or output rpm. The percentage of power going out the carrier is

$$\frac{HP \text{ primary carrier}}{HP_{in}} = \frac{(R_{s1} + R_{r1}) n_c}{R_{s1} N_{s1}} = \frac{1}{R} \left(1 + \frac{R_{r1}}{R_{s1}}\right)$$

The exact percentage of total power transmitted directly by the primary carrier is dependent on the primary- and secondary-stage ratio split but is approximately one-third of the total. The remaining two-thirds of the power is transmitted through the secondary stage. The power split-up serves to reduce secondary-stage loading and increases gear efficiency. The disadvantage of the split power system is that it is mechanically complex.

Figure 9.9 shows a different split power variation which is used to achieve ratios in the range 24:1 to 55:1. The application here is a wheelmotor which combines a hydraulic motor and the planetary gearing as a single unit in the wheel hub. Wheelmotors are used in mobile construction and off-the-road equipment where a centrally located pump drives the hydraulic motors at each wheel.

Figure 9.10 illustrates a wheelmotor casing. In the arrangement of Figure 9.9 the hydraulic motor drives through a sun gear. The output members are the primary- and secondary-stage ring gears which are connected to the wheel rim. The power split up is as follows: Part of the input goes directly from the primary

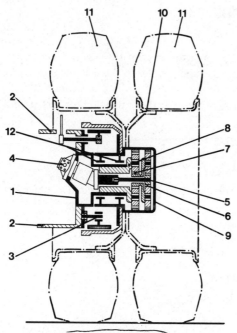

Figure 9.9 Wheelmotor transmission. Key: 1, casing; 2, vehicle chasis; 3, road brake; 4, hydraulic motor; 5, drive shaft; 6, first planetary stage; 7, second planetary stage; 8, axle journal; 9, hub; 10, rim; 11, twin tires; 12, roller bearing. (Courtesy of American Lohmann Corporation, Hillside, N.J.)

sun to the primary ring gear. The remainder goes directly from the primary carrier to the secondary sun gear and then to the secondary ring gear.

Power Feedback Systems

When working with complicated gear systems it is possible to arrange the gears in such a manner that the power transmitted by some components is greater than the input power. Although these arrangements generally achieve unusually large ratios, the component size must be sufficient to handle the recirculating power and this disadvantage may offset the large ratio obtained.

  If it is not recognized that a system has recirculating power, failures may occur. Figure 9.11 illustrates a feedback system. The input is to the primary-stage sun gear, which is connected to the secondary-stage sun gear. The output is the secondary-stage carrier. Following are the speed equations for this system:

Figure 9.10 Wheelmotor casing. (Courtesy of American Lohmann Corporation, Hillside, N.J.)

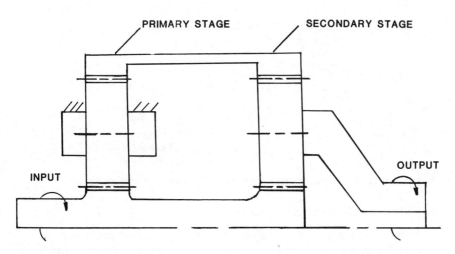

Figure 9.11 Feedback system.

$$W_{s1} R_{s1} = W_{r1} R_{r1} = W_{p1} R_{p1}$$

$$W_{r2} R_{r2} = W_{p2} R_{p2} - W_{c2} R_{r2}$$

$$W_{s2} R_{s2} = W_{p2} R_{p2} + W_{c2} R_{s2}$$

$$W_{s2} = W_{s1}$$

$$W_{r2} = W_{r1}$$

Combining the equations above the speed ratio is

$$R = \frac{W_{s1}}{W_{c2}} = \frac{1 + R_{r2}/R_{s2}}{1 - \dfrac{R_{s1} R_{r2}}{R_{r1} R_{s2}}}$$

Let us assume the following dimensions:

$$R_{s1} = 1 \text{ in.}$$

$$R_{s2} = 1.05 \text{ in.}$$

$$R_{r1} = 8.3 \text{ in.}$$

$$R_{r2} = 8.25 \text{ in.}$$

The reduction ratio $R = 166{:}1$. Let us determine the horsepower transmitted:

$$\text{hp out} = 2W_{TS2} \frac{R_{s2} + R_{r2}}{2} \frac{n_{c2}}{63,025}$$

where

W_{TS2} = secondary-stage sun tangential load, lb
n_{c2} = carrier rpm

$$\text{hp secondary sun} = \frac{W_{TS2} R_{s2} N_{s2}}{63,025}$$

where N_{s2} is the secondary sun in rpm.

$$\frac{\text{hp secondary sun}}{\text{hp out}} = \left(\frac{R_{s2}}{R_{s2} + R_{r2}} \right) R = 18.74$$

The power circulating in the gear train is 18.74 times the output power.

Planetary Gear Design Considerations

Although the basic gear tooth design of planetary gear configurations is no different from parallel shaft gearing, there are several points that must be considered when rating planetary gears and defining the detail geometry. This section discusses load sharing, assembly, and choice of numbers of teeth and numbers of planets.

Load Sharing

The main advantage of planetary gearing, of course, is transmittal of load through two or more parallel paths. Ideally, the load should be shared equally by each planet gear, but due to manufacturing errors this is never the case. Some of the factors affecting load sharing are:

Inaccurate carrier bore location
Inaccurate gear tooth spacing
Variation in planet tooth thickness
Eccentricities between sun, ring, and carrier

Figure 9.12A shows the load distribution of a three-planet system. The situation is such that in order for the sun gear to be in static equilibrium the planets must share load equally. If the sun gear is free to move and one planet is overloaded, the sun gear will shift to equalize the planet loads. Figure 9.12B shows static equilibrium; however, the load is not equally shared between planets.

In general, there are three methods of attacking the problem of load sharing between planets:

1. A completely rigid system which relies on precise component tolerances
2. A system with flexibility built in by floating one or more of the members
3. Systems that rely on mechanical means to adjust the planets to provide load equalization

The most straightforward approach is a combination of 1 and 2, where tolerances are closely held and the sun or ring or both are flexible to adjust to load maldistributions. This technique is used in high-speed planetaries such as those in gas turbine engines. Strain gage studies have shown that load sharing within 10% can be practically achieved.

AGMA Standard 420.04, Practice for Enclosed Speed Reducers or Increasers Using Spur, Helical, Herringbone and Spiral Bevel Gears, states the following:

> To compensate for unequal loading of multiple planet pinions, the total power capacity of all pinions should be the calculated capacity of one pinion plus a maximum of 0.9 times the calculated capacity for each additional pinion. For two planet pinions multiply the calculated capacity of one pinion by 1.9; for three pinions multiply by 2.8. If a load balancing device is used which insures equal loading of all planets, the capacity modifying factors need not be used.

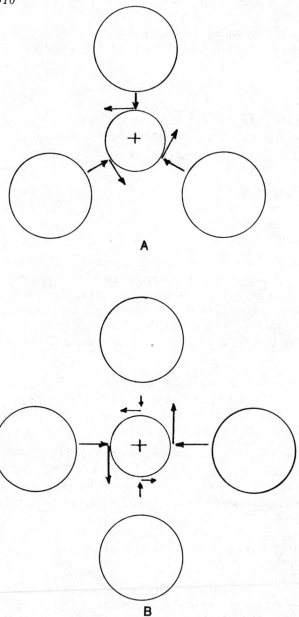

Figure 9.12 Planetary load distribution. A. Three planet system–load must be equally shared if sun is free to float. B. Four planet system in static equilibrium with unequal load distribution.

For any specific application the gear designer must determine what degree of load sharing is anticipated and rate the gearset accordingly. This estimate may be based on analysis, test, or experience, but it is important to understand that perfectly equal load sharing will not be achieved.

STOECKICHT DESIGN

A planetary configuration, extensively used in Europe for many years, is commonly called the Stoeckicht design, named after the inventor. Figure 9.13 illustrates this double helical configuration, which features flexibility not only in the sun gear but also in the ring gears, which are connected to the housing by a series of splines. The purpose of the splines is to achieve load equalization not only between the planets but also between each of the two helical gears. This system has been very successful in the past since the many degrees of freedom tend to compensate for tooth errors. There are disadvantages to the Stoeckicht design:

1. Use of through-hardened double helical gearing does not achieve as small an envelope as hardened and ground single helical gearing.
2. The design is mechanically complex.
3. The design does not lend itself to compounding to achieve higher ratios.

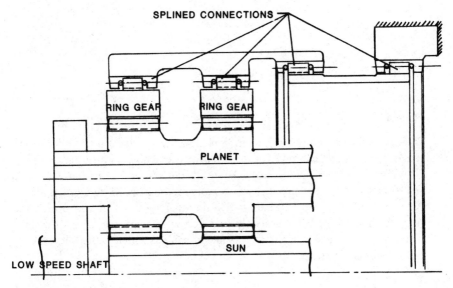

Figure 9.13 Stoeckicht planetary system design.

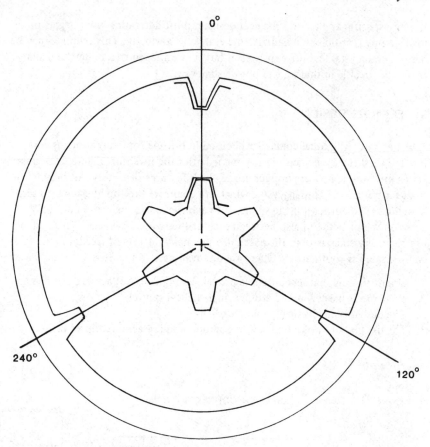

Figure 9.14 Simple planetary gear assembly.

Planetary Gear Assembly

When choosing the numbers of teeth in planetary gears, it must be understood that not every combination of gear teeth can be assembled. For instance, Figure 9.14 shows a three-planet set with a six-tooth sun gear and an 18-tooth ring gear. In this instance it is obvious that planet gears can be placed into mesh at the 0°, 120°, and 240° positions and there is no assembly problem. However, if the ring gear had 19 teeth the gearset would not be assembleable. For a simple planet set (sun, planet, ring) to be capable of being assembled, the following equation must be satisfied:

$$H = \frac{N_s + N_r}{K}$$

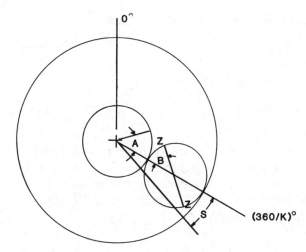

Figure 9.15 Simple planetary gear assembly analysis.

where

H = a whole number (integer)
N_s = number of sun gear teeth
N_r = number of ring gear teeth
K = number of planets

To understand this equation, refer to Figure 9.15. Assume a three-planet system with a planet gear located at the $0°$ position as shown. In order to assemble a planet at the $120°$ position, the centerline of a tooth or tooth space of the sun gear must arrive at the $120°$ point at the same time as the centerline of a tooth or tooth space of the ring gear. Let A be the angle between the center of that sun gear tooth or space and the $120°$ line and S be the angle between that ring gear tooth or space and the $120°$ mark. The following equation can be written for A:

$$A = \frac{360}{K} - \frac{360H_s}{N_s}$$

where

K = number of planets
H_s = whole number of sun gear teeth
N_s = total number of sun gear teeth

Note that the equation is general and can be applied to systems with any number of planets.

A similar equation can be written for the ring gear angle S:

$$S = \left(\frac{360}{N_r} \quad H_r\right) - \frac{360}{K}$$

where

H_r = whole number of ring gear teeth
N_r = total number of ring gear teeth

Line ZZ on the planet (Figure 9.15) passes through either two tooth centers or tooth spaces or one tooth center and one tooth space, depending on whether the planet has an even or odd number of teeth. The planet rotates through an angle B at the same time the sun rotates through angle A and the ring gear through angle S. The relationship between these angles is

$$B = \frac{AN_s}{N_p} = \frac{SN_r}{N_p}$$

where N_p is the total number of planet teeth. Substituting for A and S yields

$$\left(\frac{360}{K} - \frac{360H_s}{N_s}\right)\frac{N_s}{N_p} = \left(\frac{360H_r}{N_r} - \frac{360}{K}\right)\frac{N_r}{N_p}$$

which simplifies to the assembly equation

$$H = \frac{N_s + N_r}{K}$$

The assembly equation for a compound planet system is more complicated and can be understood by considering Figure 9.16. Assume that points M and N are registry teeth on the compound planet. By this is meant that the centerline of a tooth on the primary planet is aligned with the centerline of a secondary planet tooth. The purpose of the registry marks is to enable all compound planets to be manufactured identically.

As was shown in the analysis of the simple planet system, the center of a sun gear space rotated an angle A from the 120° mark must arrive at the 120° mark at the same time as the center of a ring gear space, which is shown rotated an angle S from the 120° tooth mark.

$$A = \frac{360}{K} - \frac{360H_s}{N_s}$$

$$S = \frac{360H_r}{N_r} - \frac{360}{K}$$

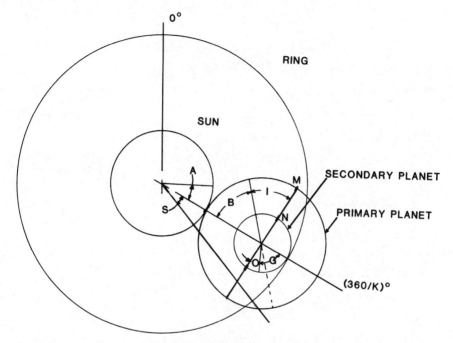

Figure 9.16 Compound planetary gear assembly analysis.

The primary planet rotates through an angle B at the same time the sun rotates through A:

$$B = \frac{AN_s}{N_{p1}}$$

where N_{p1} is the total number of primary planet teeth. The secondary planet rotates through an angle G at the same time the ring gear rotates through S.

$$S = \frac{GN_r}{N_{p2}}$$

where N_{p2} is the total number of secondary planet teeth.

The relationships between the registry teeth and the meshing teeth on the primary and secondary planets are:

$$I = \frac{360H_{p1}}{N_{p1}}$$

$$O = \frac{360N_{p2}}{N_{p2}}$$

where

H_{p_1} = whole number of primary planet teeth
H_{p_2} = whole number of secondary planet teeth

From Figure 9.16,

$$I = G + O - B$$

Substituting for I, G, O, and B, we have

$$\frac{360H_{p_1}}{N_{p_1}} = \frac{N_r}{N_{p_2}}\left(\frac{360H_r}{N_r} - \frac{360}{K}\right) + \frac{360H_{p_2}}{N_{p_2}} - \frac{N_s}{N_{p_1}}\left(\frac{360}{K} - \frac{360H_s}{N_s}\right)$$

and simplifying yields

$$\frac{N_r + N_sN_{p_2}/N_{p_1}}{K} + (H_{p_1} - H_s)\frac{N_{p_2}}{N_{p_1}} = H_r + H_{p_2}$$

In order to solve this equation the term on the left,

$$\frac{N_r + N_sN_{p_2}/N_{p_1}}{K}$$

is calculated and whole numbers are substituted for (H_{p_1} - H_s). If the left side of the equation works out to be a whole number, the gearset can be assembled.

In a compound planetary each of the planets must be identical not only for assembly purposes but also for proper load sharing. The registry teeth are used to satisfy this requirement and a typical allowable tolerance for the variation of the centerline of the primary planet tooth with the centerline of the secondary tooth is ±0.001 in. Figure 9.17 is a photograph of a compound planetary gear. Compound planets can be either of two-piece or one-piece construction. The processing must be very precise to achieve the registry dimension. In addition to the registry, the tooth thickness and stock removal must be closely controlled. The sequence of operations for a two-piece hardened and ground compound planet are as follows:

1. Cut primary and secondary planet teeth
2. Heat treat
3. Finish grind secondary planet
4. Press primary onto secondary using a fixture to align registry teeth
5. Grind primary planet to final tooth size and registry dimension

The reverted compound planetary gearset (Figure 9.6) with an input and output sun gear has the following assembly equation:

$$\frac{N_0 - N_{s_1}N_{p_2}/N_{p_1}}{K} + (H_s + H_{p_1})\frac{N_{p_2}}{N_{p_1}} = H_{p_2} + H_0$$

Figure 9.17 Compound planet being balanced.

where

N_0 = total number of low-speed sun gear teeth
H_0 = whole number of low-speed sun gear teeth

It should be noted that it may be physically possible to assemble a gearset that does not satisfy the assembly equations or has mistimed registry markings. If there is sufficient backlash it may be possible to force the components into mesh; however, the gears at some point in the rotation will tend to bind and will have poor load sharing.

Choice of Tooth Numbers

There are three considerations in choosing the numbers of teeth of a planetary in addition to the basic problem of achieving the correct ratio:

1. Assembly considerations as discussed in the preceding section
2. Choosing hunting teeth
3. Achieving sequential mesh rather than simultaneous mesh

By hunting teeth it is meant that the numbers of teeth in a gear mesh be such that each tooth in the pinion at some point in time mesh with each tooth in the gear. For instance with a 20-tooth sun and 40-tooth planet each sun gear tooth would always mesh with the same 2 planet teeth. By changing the number of planet teeth to 41, a hunting tooth mesh is achieved.

The advantage of having a hunting tooth is that the gearset has more of a tendency to wear in. For instance, if one tooth has a high spot and meshes with all the mating gear teeth, there is a better chance that the error will wear in and not cause permanent damage than if the discrepant tooth kept mating with the same two meshing gear teeth.

Sequential mesh occurs in a sun-planet mesh when the number of teeth in the sun is not divisible by the number of planets. When the number of teeth in the sun is divisible by the numbers of planets the mesh is called simultaneous. In this case each sun-planet mesh is occurring at the same point on the gear at the same time. For instance, when the planet at the $0°$ point is in mesh at its pitch line, the $120°$ and $240°$ planets in a three-planet system are also being loaded at the pitch line. This situation should be avoided, if possible, since the load impulse the simultaneous mesh gives to the mechanical system is stronger than that of the sequential mesh. The sequential mesh, therefore, is less likely to generate harmful vibrations.

The mesh frequency of a simple gearset is the number of teeth times the speed. If a planetary gear has sequential meshing, the mesh frequency will be the number of planets times the number of teeth times the speed. The higher sequential mesh frequency is less likely to excite any natural frequencies in the operating equipment.

Although having hunting teeth and sequential mesh is desirable, it is not always achieved and many gearsets operate successfully without fulfilling these conditions. The most obvious way of determining the number of planet teeth is to subtract the number of sun teeth from the number of ring teeth and divide by 2. Let us take an example:

N_s = number of sun gear teeth   = 20

N_r = number of ring gear teeth   = 100

CD = center distance, inches   = 10

The planet gear geometry is

$$N_p = \frac{N_r - N_s}{2} = 40$$

$$DP = \frac{N_p + N_s}{2(CD)} = \frac{N_r - N_p}{2(CD)} = 3.0$$

where DP is the diametral pitch.

$$PD_p = \frac{N_p}{DP} = 13.3333 \text{ in.}$$

What if the number of ring gear teeth minus the number of sun gear teeth is not even? Let us assume the following case:

$$N_s = 20$$

$$N_r = 103$$

$$CD = 10$$

$$N_p = \frac{N_r - N_s}{2} = 41.5$$

If we make the number of planet teeth 41 there is no problem in achieving a practical gear set, but the operating conditions are different at the sun-planet mesh from the planet-ring mesh. At the sun-planet mesh:

$$DP_{s-p} = \frac{N_s + N_p}{2(CD)} = 3.05$$

The planet pitch diameter is

$$PD_{p(s-p)} = \frac{N_p}{DP_{s-p}} = 13.4426$$

At the planet-ring mesh:

$$DP_{p-r} = \frac{N_r - N_p}{2(CD)} = 3.10$$

$$PD_{p(p-r)} = \frac{N}{DP_{p-r}} = 13.2258$$

If the pressure angle ϕ is chosen at $22.5°$ at the sun-planet mesh, the pressure angle at the planet-ring mesh is calculated as follows:

$$\cos \phi_{s-p} \cdot PD_{p(s-p)} = \cos \phi_{p-r} \cdot PD_{p(p-r)}$$

and $\phi_{p-r} = 20.1117°$. The lead of the planet gear is

$$L = \frac{\pi \cdot PD}{\tan \psi}$$

If the helix angle ψ is chosen as $10°$ at the sun-planet mesh, the helix angle at the planet-ring mesh is as follows:

$$\tan \psi_{(p\text{-}r)} = \tan \psi_{(s\text{-}p)} \cdot \frac{PD_{p(p\text{-}r)}}{PD_{p(p\text{-}s)}}$$

and $\psi_{p\text{-}r}$ = 9.8419°. The fact that the operating conditions at the sun-planet mesh are different from those at the planet-ring mesh is advantageous. The sun-planet mesh is critical in terms of compressive stress and flash temperature rise; therefore, it is desirable to have a large pressure angle which reduces both these parameters. Because the ring-planet mesh is internal, compressive stress and flash temperature rise are not usually a problem and the lower pressure angle can be tolerated. In fact, a lower pressure angle results in lower separating forces on the ring gear and this is desirable to reduce ring distortion and stress under load. Even in the case where the number of ring gear teeth minus the number of sun gear teeth is an even number, it can be advantageous to drop one or two teeth from the planet in order to achieve a larger pressure angle at the sun-planet mesh pitch diameter and a smaller pressure angle at the planet-ring mesh pitch diameter.

The question may be asked: How many planets should one use in a particular planetary set? There is a physical limit to the number of planets that can fit into any specific application. Figure 9.18 plots the maximum number of planet gears which can be assembled into a planet system versus the ratio of the sun gear diameter divided by the planet gear diameter. As the reduction ratio becomes smaller, the sun gear becomes larger and more planets can be fit in. As shown earlier, in a three-planet system with a floating sun gear, the load between the planets tends to be equalized. As the number of planets is increased, load sharing may suffer and the full benefit of the additional planets may not be realized. The vast majority of planetary gears use three planets; however, systems with 4, 5, 7, and more planets have been successfully developed.

A point that must be considered when planet gears are idlers meshing with both the sun and ring gear is that the planet teeth experience complete reversal of stress. Allowable bending stresses must be multiplied by 0.7 to compensate for this loading condition.

Planetary Bearing Loads

Quite often the critical components from a life point of view in a planetary system are the bearings. Because the planet gears are loaded both at the sun and ring gear meshing points (Figure 9.4), the planet bearings must react twice the tooth loads.

In an epicyclic system where the planet carrier rotates about the center of the system, the centrifugal force on the planet gears must be reacted by the planet bearings. The centrifugal force on a planet gear is calculated as follows:

$$F_c = mrW^2$$

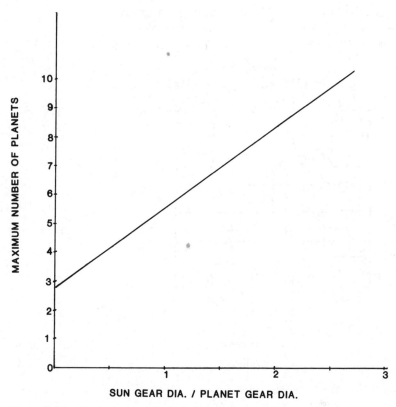

Figure 9.18 Maximum number of planets that can be assembled.

where

F_c = centrifugal force, lb
m = planet mass, lb-sec^2/in.
r = radius to planet center, in.
W = carrier angular velocity, rad/sec

Let us work through an example of a helical epicyclic planetary gear set and calculate the planet bearing loads.

PD_s = sun gear pitch diameter, in. = 3.5

PD_p = planet gear pitch diameter, in. = 3.5

HP = transmitted horsepower = 1000

n_s = sun rpm = 7200

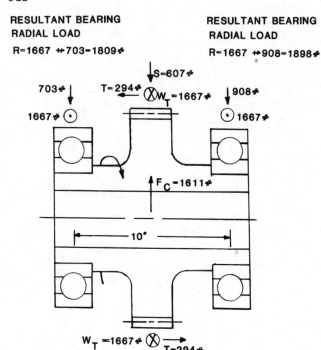

RESULTANT BEARING
RADIAL LOAD
R=1667 ++703=1809#

RESULTANT BEARING
RADIAL LOAD
R=1667 ++908=1898#

Figure 9.19 Planet bearing loading.

n_c = carrier rpm   = 1800

ψ = helix angle, deg   = 10

ϕ = pressure angle, deg   = 20

Figure 9.19 illustrates the planet configuration and loading. Assume that the planet weighs 5 lb. The centrifugal force is

$$F_c = \frac{5}{386}\ (3.5)\left(\frac{1800\ \cdot\ 2\pi}{60}\right)^2 = 1611\ \text{lb}$$

The tangential load per planet with a three-planet system is

$$W_T = \frac{63,025(1000)}{7,200}\left(\frac{2}{3.5}\right)\frac{1}{3} = 1667\ \text{lb}$$

The separating load at each mesh is

$$S = W \tan \phi = 607\ \text{lb}$$

The thrust load at each mesh is

$$T = W_T \tan \psi = 294 \, \text{lb}$$

For the planet to be in static equilibrium the bearing loads are as shown in Figure 9.19. It can be seen that the centrifugal load is significant. The limiting factor in an epicyclic system quite often is the carrier speed, which, if excessive, results in overloading of the planet bearings due to centrifugal force.

Planetary Gear Economics

At first glance it would appear that a planetary gearset must be more expensive than a parallel shaft configuration. There is the added complication of a carrier and two or more additional planet gears plus more bearings. Ring gears are usually expensive items.

In fact, the reduced size of planetary components offsets the cost of additional parts and the determination of which design is less costly is not so obvious. An important factor is the quantity to be manufactured and as the quantity of units increases, the savings in material and advantages of handling and machining smaller components begin to outweigh the disadvantages of mechanical complexity.

Also, when high ratios are required (over 15:1) planetaries tend to be more economical, even in relatively small quantities.

In very high horsepower units, the components in parallel shaft boxes tend to become so large that there may be an economic advantage to planetaries even in the ratio range 2:1 to 15:1.

In any case, the technical advantages of planetary gearing are sufficient such that this type of gearing should be considered, especially since the economic penalty may be small or nonexistent.

10

GEARBOX INSTALLATION:
MOUNTING, ALIGNMENT, COUPLINGS

Proper installation of the geared system is essential to achieve good performance. The gearbox must be rigidly connected to the foundation, which must also be rigid and have a flat mounting surface. If the foundation or base plate structure is incorrectly designed or constructed vibration, shaft misalignment, bearing damage, and even shaft or housing breakage can result.

Most gearboxes are foot mounted. A flange on the bottom of the unit is doweled and bolted to a base. The doweling is important to ensure that the gearbox does not move during operation. This is a four-point mounting arrangement, and since three points define a plane it is difficult to install the unit such that all four points lie in the same plane. Two major reasons why the mounting points on the base plate are usually not coplanar are:

1. The steelwork warps as a result of poor welding, grouting, or concrete work.
2. The use of multiple steel beams which are not coplanar in the base plate.

When mounting the unit on multiple steel beams a base plate which extends under the entire gearbox and is at least as thick as the gearbox base should be used. Both the gearbox and the base plate should be rigidly bolted to the steel supports.

Shims can be used to bring all mounting surfaces into the same plane. Care must be taken that the shims form a solid tight pack when the bolts are tightened. Prior to final tightening, the shims should be inspected for rust, folds, wrinkles, burrs, tool marks, and dirt. Correct grouting is of great importance when bases are supported on concrete. Figure 10.1 is a checklist to ensure proper grouting.

Proper grouting checklist

1. Use non-shrink grout, as confirmed by test data per ASTM C-827. Volume change after hardening should be the measurement used.

2. Make sure baseplate design permits a complete fill, with adequate pour openings and inspection holes.

3. If epoxy grout is to be used, ask for evidence that long-term cold flow or "compression creep" will not occur.

4. Avoid metallic aggregate grouts where either severe temperature swings or corrosive surroundings may exist.

5. Keep the concrete foundation moist for 24 to 48 hours prior to grouting.

6. Be sure all oil, grease, or dirt is removed from any surface to be contacted by grout.

7. Place grout continuously, as quickly as possible, and work from one side or end across to the other side or end, to avoid trapping air in confined spaces.

8. Ensure that the surrounding temperature and the chemical additives allow ample working time for proper placement before setup (at least 45 minutes for a full foundation).

9. Allow enough curing time before placing the drive in operation; depending on the type of grout, this could be as long as 7 days.

10. When the drive is in service, watch for signs of deterioration of grout.

One result of ignoring the checklist above.

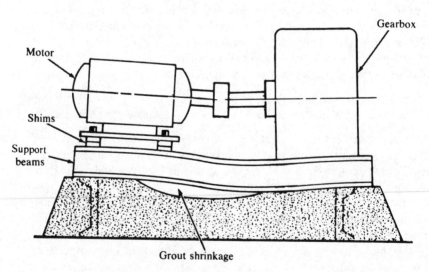

Figure 10.1   Grouting procedures. (From Ref. 1.)

Stability of the foundation is extremely important. Once installed, the base must not deform. One major cause of difficulty is thermal expansion, which can be due either to ambient conditions such as partial sunshine unevenly heating the structure or proximity to hot operating equipment. Of course, the structure must initially be designed with sufficient rigidity to withstand all operating forces without distortions.

It is possible to encounter a resonance condition where the natural frequency of the complete system assembly, including the base plate, coincides with an operating frequency. This type of structural resonance can be corrected in several ways:

1. The natural frequency of the base can be increased by making it more rigid by adding stiffeners or gussets.
2. The natural frequency of the base can be lowered by removing material from the base.
3. The mounting system can be made more elastic to dampen or lower the frequency. An example of this is the use of spring washers at the attachment points.
4. The mass of the rotating system can be changed to modify its natural frequencies.

When handling the unit at installation, care must be taken not to stress parts which are not meant to support the gearbox weight. Gearboxes should be lifted only by the means provided by the manufacturers, such as lifting holes in the casing.

COUPLINGS AND SYSTEM ALIGNMENT

To connect the driving and driven equipment to the gearbox, input and output shaft couplings are used. It would not be practical to align the centerlines of the equipment exactly; therefore, the coupling must have some degree of flexibility to accommodate misalignment. Even if it were possible to perfectly align the equipment at some given operating point, because load, speed, or ambient conditions vary, the alignment will change. Figure 10.2 illustrates the alignment conditions that the equipment shaft ends may be in. The coupling must accommodate these conditions while transmitting torque, yet limit the forces on machine components such as shafts and bearings that result from misalignment. If the coupling had no flexibility, the misalignment would have to be accommodated by bending of the shafting, which would apply loads to the bearings. The maximum amount of misalignment couplings can accommodate must be satisfied at assembly with alignment procedures that will be described below. It is important, however, to attempt to minimize the misalignment to the lowest practical values since coupling life is strongly dependent on how well the system is aligned.

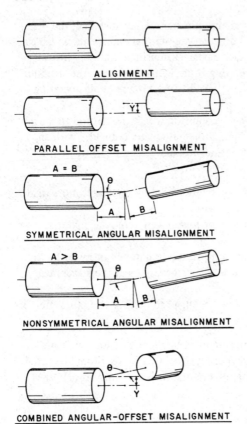

ALIGNMENT

PARALLEL OFFSET MISALIGNMENT

SYMMETRICAL ANGULAR MISALIGNMENT

NONSYMMETRICAL ANGULAR MISALIGNMENT

COMBINED ANGULAR-OFFSET MISALIGNMENT

Figure 10.2 Shaft alignment conditions. (From Ref. 2, courtesy of American Gear Manufacturers Association, Arlington, Va.)

Figure 10.3 illustrates the most widely used coupling types. With gear couplings, misalignment is accommodated by sliding of the spline teeth. The sliding results in wear, and excessive wear is the major cause of failure of this type of coupling. In order to limit wear, gear couplings are lubricated, usually with grease. The maximum sliding velocity is [3]

$$V_s = \frac{D}{2} \; \theta \; \frac{2\pi \cdot \text{rpm}}{60}$$

where

V_s = sliding velocity, ips
θ = misalignment angle, in./in. (rad)
D = pitch diameter, in.

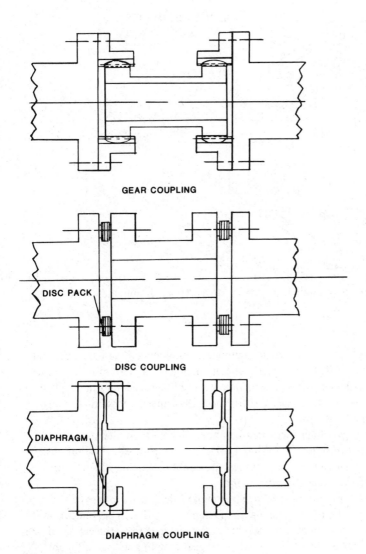

Figure 10.3 Coupling types.

Reference 3 gives the following guidance for allowable sliding velocities:

Less than 1.4 ips: optimum condition
3.0 ips: normal operation
5.0 ips: maximum allowed

Gear couplings can accommodate axial motion by allowing the male splines to float axially within the female splines as shown in Figure 10.3.

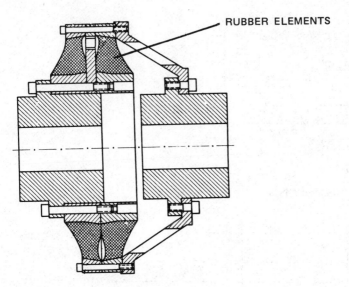

Figure 10.4 Elastic coupling. (Courtesy of American Lohmann Corporation, Hillside, N.J.)

Disk and diaphragm couplings accommodate misalignment and axial motion by flexing of the elastic elements. These couplings generally fail in fatigue and the fatigue life is dependent on misalignment, coupling speed, and the applied steady-state loads due to torque transference. Disk and diaphragm couplings require no lubrication.

Elastic couplings which incorporate rubber elements bonded to a steel backing are sometimes used with diesel drives to damp torsional vibrations. Figure 10.4 illustrates an elastic coupling with rubber elements which dampen vibration and accommodate misalignment.

The maximum amount of angular misalignment flexible couplings can accommodate is in the order of ¼°. The maximum allowable offset is dependent on the distance from pivot to pivot. For instance, if the distance from the center of one spline to the center of the other spline on the gear coupling shown in Figure 10.3 is 10.0 in. and ¼° or 0.0044 in./in. maximum angular misalignment is possible, the maximum inches of parallel offset are

0.0044(10.0) = 0.044 in.

The values above are the maximum flexible couplings can accommodate; however, operation at these limits will shorten coupling life and may be detrimental to the system. During operation misalignment should not exceed approximately one-fifth of the foregoing values and the system should be carefully aligned prior to startup to achieve this.

ALIGNMENT PROCEDURE

An outline of the steps required to align two pieces of equipment connected by a coupling is as follows:

1. Roughly set up the machines visually and with crude measurements.
2. Accurately align the shafts in the cold condition using precise measurements.
3. Operate the equipment and accurately measure the alignment in the hot condition.
4. Calculate the cold alignment required to achieve alignment in the hot condition. For instance, if it is determined that a motor shaft rises 0.005 in. with respect to the gearbox shaft during operation, the motor shaft should be positioned 0.005 in. below the gearbox shaft during cold alignment.
5. Realign in the calculated cold condition, operate the system, and recheck alignment in the hot condition.

The alignment procedure begins with the coupling hubs mounted on their respective shafts. The equipment is placed in position with the proper axial gap X, as shown on Figure 10.5. In order to roughly align the shafts, crude measurements can be made with a straightedge and calipers. The straightedge checked in

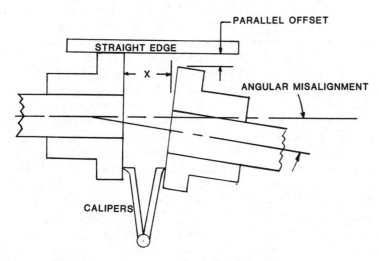

Figure 10.5 Rough alignment of coupling halves.

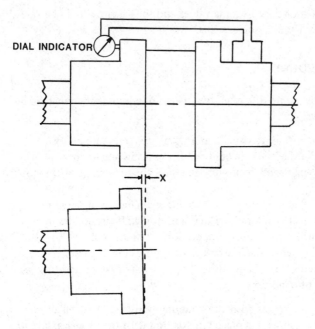

Figure 10.6 Checking for angular misalignment.

two planes will give an indication of the parallel offset. Caliper measurements at four points will give an indication of the angular misalignment.

There are several methods available to accurately align the shaft ends. One method, using a dial indicator, will be described. The first step, as shown in Figure 10.6 is to check for angular misalignment. A dial indicator base is mounted securely on the right-hand hub and the dial indicator stem is placed against a face on the left-hand hub. The connected shafts are rotated several times and the dial indicator checked to ensure ample movement in either direction. It is useful to use a mirror to observe the dial indicator gage as the shaft is rotated. The point at which a minimum reading is registered is found and at this point the dial indicator gage is set at zero. When the shaft is rotated 180° from this point the dial indicator reading will be the total angular misalignment. As an example, refer to Figure 10.6. If the minimum reading occurred at the bottom position, when the coupling is rotated one-half turn and the dial indicator is at the top position as pictured, the indicator will read the dimension X, which means that the distance between the coupling faces is X inches greater at the top than at the bottom. In order to align the faces, the equipment must be shimmed to narrow the distance between the faces at the top to X/2 inches. The distance between the faces at the bottom will increase X/2 inches and the faces will be square.

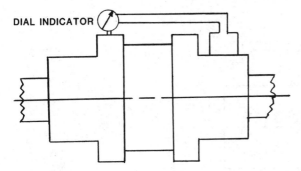

Figure 10.7 Checking for parallel offset.

Now that the angular misalignment is corrected, parallel offset must be measured. Figure 10.7 shows the dial indicator base securely fastened to the right-hand coupling hub with the stem in contact with a smooth outside diameter on the left-hand hub. Again the shafts are rotated several turns, making sure the indicator has travel in both directions and the point where the reading is minimum is found. The dial indicator is set at zero at this point and the shaft rotated 180°. In this position the indicator will read twice the amount of parallel offset. For instance, in Figure 10.7 if the minimum indicator reading was at the bottom position, when the coupling is rotated one-half turn and the indicator stem is at the top position, as pictured, if the reading is 0.050 in. the parallel offset of the center lines of the shaft ends is 0.025 in. with the left-hand shaft higher than the right-hand shaft. The equipment can now be shimmed vertically by 0.025 in. to bring the coupling into alignment.

After correcting for parallel offset the axial spacing and angular measurement should be rechecked to make sure that they were not disturbed. If the bracket holding the dial indicator is not rigid and allows the indicator to sag, an error will be introduced into the alignment readings. Readings in the horizontal plane will be little affected, but in the vertical plane, sag will increase the reading when the indicator is on top and decrease the reading when the indicator is on the bottom. One way to measure if the dial indicator is sagging is to mount the breacket on an accurately machined cylinder. The top reading, T (Figure 10.8), is greater than the nominal reading by the amount of sag and the bottom reading, B, is less than the nominal reading by the amount of sag. The sag is

$$\text{Sag} = \frac{T}{B}$$

The parallel offset determined by the method shown in Figure 10.7 must be corrected for the sag. This can be done by lowering the centerline of the measured shaft by the amount of sag. For instance, if the centerline of the left-hand shaft is measured to be 0.025 in. higher than the right-hand shaft and the

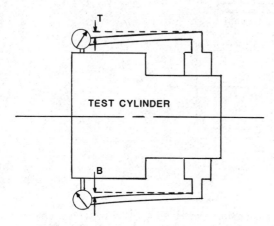

Figure 10.8   Measuring dial indicator bracket sag.

sag is measured to be 0.005 in., the actual parallel offset in the vertical plane is 0.020 in.

In Figures 10.6 and 10.7 both shafts are rotated when making the angular misalignment and parallel offset measurements. It would be possible to rotate only the shaft holding the dial indicator bracket and traverse the face and outside diameter of the other shaft while it is stationary. In this case any outside diameter eccentricity or face runout will be a source of error in the alignment reading. This problem is averted by rotating the measured surface together with the indicator.

Dial indicator methods of measuring misalignment have the disadvantage of not being able to monitor machine condition while operating. There are optical and proximity probe systems which can be used to monitor alignment continuously when an application warrants the expense of such instrumentation [3].

REFERENCES

1. Nailen, R. N., Trouble Free Drives Are Based on Firm Foundations, *Power Transmission Design*, May 1981, pp. 38–42.
2. AGMA Standard 510.02, Nomenclature for Flexible Couplings, American Gear Manufacturers Association, Arlington, Va., August 1969.
3. Cox, J. L. and Wilde, L. G., Alignment of Turbomachinery, *Sawyers Turbo-Maintenance Handbook*, Vol. III, Turbomachinery International Publications, Norwalk, Conn., 1980.

BIBLIOGRAPHY

Flexible Coupling Two Step Dial Indicator Method, Rexnord Inc., Coupling Division, Warren, Pa.

11

GEAR UNIT OPERATION: TESTING, STARTUP, CONDITION MONITORING

In this chapter three phases of gearbox operation are discussed: testing, initial field startup, and condition monitoring.

TESTING

During the procurement process the gearbox manufacturer and the user must agree on the type of testing the completed unit will be subjected to prior to acceptance. The test program can be as simple as turning the shafting by hand to verify free operation of the internal components or as complicated as full-scale operation of an instrumented gearbox, on the actual application, through a predetermined test schedule.

Several factors must be considered when determining the degree of test program complexity:

1. Cost of testing
2. Confidence in the gearbox design
3. Consequences of gearbox failure

Clearly, a proven gear design, operating in an environment where downtime does not incur a large cost penalty, would not warrant extensive testing. On the other hand, if a gearbox is a critical component in a complex system the cost of testing may be slight compared to the loss in case of failure. Where human safety considerations are involved, testing may be required to limit legal liability. New designs or extrapolations of existing designs warrant sufficient testing to verify the analytical and manufacturing procedures.

Test programs can be extremely valuable in identifying problem areas prior to field operation. Design, manufacturing, and assembly errors can be identified and corrected prior to a costly catastrophic failure. In order to gain the most information, the test plan and instrumentation scheme should be carefully designed and acceptance limits set prior to operation. Data should be taken over the range of speeds, loads, and operating environments anticipated in the application. To identify potential problem areas quickly, the test program should include overspeed and overload operation. Care must be taken not to go too far and develop failure modes that would not occur in the actual application, but overspeeds to 110% and overloads to 125% should be within the capability of the unit.

Some of the common problems that can be identified in initial testing are:

1. *Excessive gearbox heat generation.* This is most commonly caused by oil churning and can be corrected by improving scavenging, reducing oil flow or changing oil type.
2. *Improper gear pattern.* May be caused by gear tooth errors, bore misalignment, or deflections. Can be corrected by tooth modifications or possibly bearing bore relocation.
3. *Overheating of gears or bearings.* Usually caused by insufficient lubrication. Can be corrected by increasing oil flow, retargetting oil jets, or changing oil type.
4. *Excessive noise or vibration.* Caused by unbalance, tooth errors, assembly errors, or operating at critical frequencies.
5. *Oil leakage.* Caused by misassembly of static or dynamic seals or pressurization of gearbox cavity.

Spin Tests

In many cases a full-speed, light-load test is considered sufficient to qualify a gear unit as acceptable. A typical test program might be:

1. Operate the gearbox at maximum continuous speed until bearing and lubrication oil temperature has stabilized.
2. Increase the speed to 110% of maximum continuous speed and run for a minimum of 15 min.
3. Reduce speed to maximum continuous and run for 4 hr at a minimum.

The following measurements should be made during the acceptance test:

Oil inlet temperature
Scavenge oil temperature
Oil feed pressure
Oil flow
Shaft speed

MODEL _____   S/N _____

DATE _____   OPERATOR _____   ENGINEER _____

RUN NO.	TIME OF DAY	SPEED RPM	OIL TEMP., °F			OIL PRESSURE PSIG	OIL FLOW GPM	VIBRATION LEVEL
			1	2	3			

Figure 11.1 Typical test log sheet.

Other parameters that can be measured during gear testing are:

Vibration
Shaft excursion
Noise
Bearing temperatures

A detailed test log should be kept making entries of each measurement at regular intervals such as every 15 min. Figure 11.1 shows a typical test log.

After completion of the mechanical running test, the gear unit should be opened for a visual inspection. Tooth meshes should be inspected for surface damage and proper tooth contact. All bearings and journals should be inspected for signs of surface damage or overheating.

With high-speed gear drives it is not uncommon to conduct the full-speed acceptance test driving the gearbox through a low-speed shaft. This is done if a high-speed prime mover is not available. In this case, the gears are contacting on their normally unloaded faces since the gearbox is being driven backwards. Such a test can still be useful to determine proper operation of the lubrication system and correct alignment of the gear shafts. Also, any gross machining or assembly errors can be identified.

The full-speed "light" load test is widely used since full-load testing at a gear vendor's plant is costly and sometimes a prime mover of sufficient capacity is not available.

Load Testing

Relatively low power gearboxes (up to possibly 200 hp) are load tested with power absorption devices loading the output shaft or shafts. Power absorption

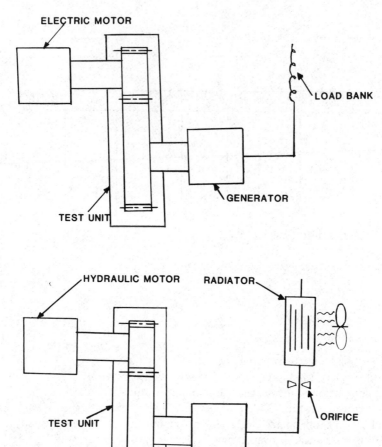

Figure 11.2   Power absorption test setups.

devices generally convert the power into heat and include water brakes, dyna-
mometers, generators with load banks, hydraulic motors driving pumps, and so
on. Prime movers include electric, hydraulic, or gasoline motors. Figure 11.2
illustrates some power absorption test setups.

For gear units transmitting high horsepower, power absorption testing is
expensive and sometimes impractical. Quite often a suitable prime mover or
power absorption device is not available to the gear manufacturer. Also, the
power required to conduct tests becomes a major expense. To test large units

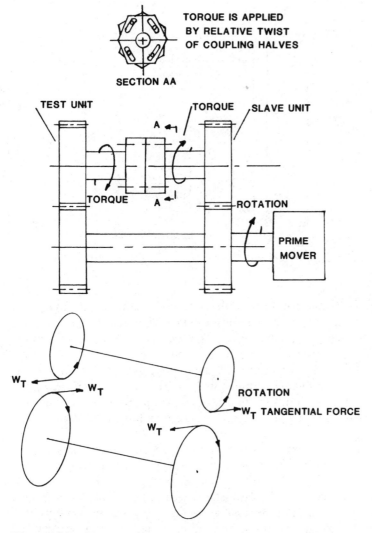

Figure 11.3 Simple regenerative rig.

at full speed and power, regenerative power techniques are used. Some other terms for this method of testing are recirculatory power, four-square rig, or back-to-back testing.

Figure 11.3 illustrates the simplest type of regenerative power test rig. Two identical gear sets are used. The high-speed shafts are coupled together, as are the low-speed shafts. Torque is applied to either the high- or low-speed shaft

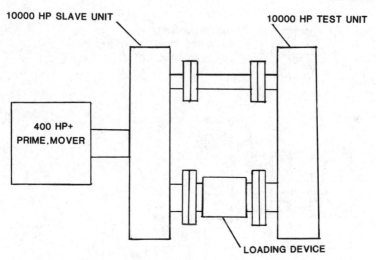

Figure 11.4   Back-to-back gear rig.

by twisting one coupling half flange with respect to the other and the torque is locked in by clamping the coupling halves together. The assembly is then rotated at the desired speed by the prime mover, which need only develop enough power to overcome the friction losses in the system.

To illustrate this point, Figure 11.4 shows two gear units set up in a back-to-back arrangement. If each unit transmits 10,000 hp and is 98% efficient the loss in the gearboxes will be 400 hp. Therefore, the prime mover for this 10,000-hp can be sized at 400 hp plus some margin for losses other than in the gear units.

With the arrangement shown in Figure 11.3, any gear tooth load desired can be developed. One gear pair is designated as the test set and the other the slave set. On the test set the torque and rotation are applied in the same sense as on the actual application. The slave set gear teeth, while loaded on the proper tooth faces, are rotating in the opposite sense. Another way to look at it is that in the test set if the pinion is driving the gear, the opposite is true in the slave set, with the gear driving the pinion. A review of the directions of rotation and torque shown in Figure 11.3 will illustrate this point.

When designing a back-to-back test thought must be given to the bearing and lubrication system operating conditions of the slave set since the opposite sense of rotation may require incorporation of some modifications. For instance, the oil feed groove in a journal bearing may have to be relocated.

The four-square type of rig gives reliable results concerning gear unit deflections under load and can be used to establish tooth modifications to

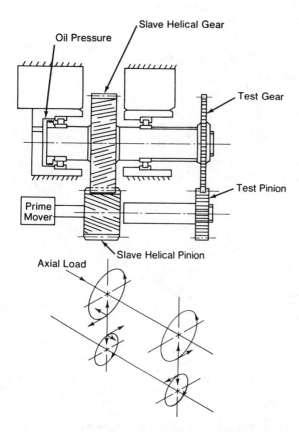

Figure 11.5 Helical gear loading regenerative rig.

improve load distribution. It is also useful in establishing efficiency values and exploring lubrication system problems. Long-term endurance testing can be accomplished at low cost.

The test setup shown in Figure 11.3 has some disadvantages:

1. The load is applied at zero speed when there is no oil film generated at the gear tooth and bearing interfaces. This may cause surface distress of the mating components. Also, it is impossible to simulate actual operating condition of most applications where load increases with speed. Starting torque requirements on the prime mover are high because full load is applied at zero speed.
2. As gears and bearings wear the locked-in load will gradually decrease.
3. The load may vary with operating temperature as components distort. This problem can be resolved by retorquing the coupling when rig temperature stabilizes.

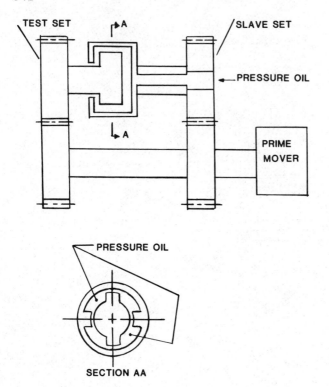

Figure 11.6 Hydraulic rotary torque actuation rig.

From these three points it can be seen that it would be advantageous to be able to apply and control the load while the rig is rotating and this can be accomplished by various means. One method shown in Figure 11.5 makes use of the fact that axial movement of a helical gear will cause rotation of the shafts. As the gear translates, all backlash is taken out of the system and torque is generated. The torque is proportional to the oil pressure applied at the load piston and therefore can be controlled at all speeds. Another method of loading a rig during operation is the rotary torque actuator shown in Figure 11.6. Pressure oil is fed to the actuator and the relative rotation of the vanes generates torque proportional to the oil pressure.

Parallel shaft units lend themselves to a four-square type of rig arrangement. Figure 11.7 shows a planetary gearbox rig where the high-speed shafts of the test and slave gearboxes are connected and operating concentrically inside the low-speed shafts. In order to apply load to the system, one of the stationary planet carriers is mounted on a loading fixture and turned to introduce a torque

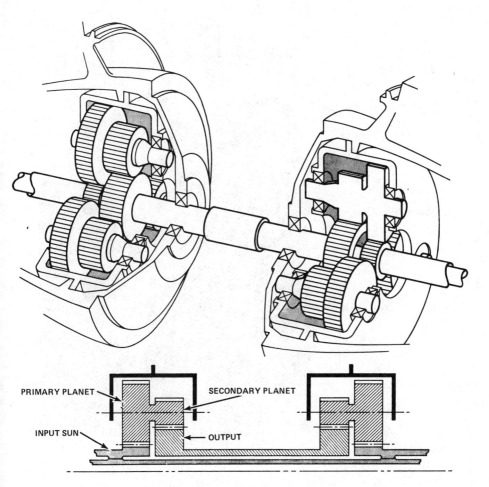

PRIMARY PLANET

SECONDARY PLANET

INPUT SUN

OUTPUT

Figure 11.7   Planetary recirculating power rig.

into the system. Figure 11.8 illustrates such a test setup with pneumatic pistons supplying a couple to turn the carrier.

Another type of recirculatory power test setup is shown in Figure 11.9. In this case only one test gearbox is required. The electric motor, through the gearbox, drives a generator, which in turn generates electricity to drive the motor. Additional power is required to offset energy losses in the system. The same technique can be used with a hydraulic motor and pump.

Static Tests

It is possible to apply full load to a stationary gearbox and determine tooth patterns and housing rigidity, as shown in Figure 11.10. The tooth patterns will

LOAD PISTON

Figure 11.8   Planetary pneumatic load device.

show up if bluing or a paste compound is applied to one gear. If a paste compound is applied, care must be taken to minimize the thickness since paste thicknesses greater than 0.0002 in. will mask the existence of misalignment between the gear teeth.

If equipment is available to conduct a low-speed, full-torque rolling test, this type of operation can be specified. For instance, a 10,000-hp, 20,000-rpm shaft can be operated at 200 rpm with a 100-hp prime mover and develop the same torque, 31,512 in.-lb. This type of test is useful to demonstrate tooth contact, load-carrying capability, and housing rigidity, but yields no information concerning parameters related to speed, such as temperature, vibration, and noise.

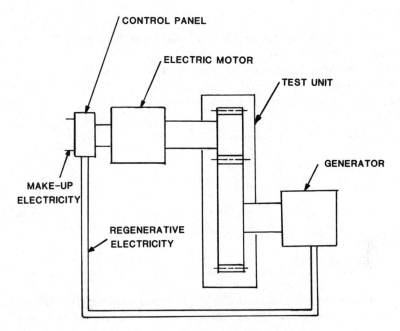

Figure 11.9   Regenerative electric motor drive rig.

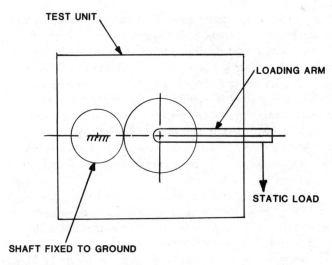

Figure 11.10   Static test rig.

Instrumentation

At a minimum during a gear test torque, speed, oil temperatures, and pressures should be monitored. Also important are vibration and oil flow.

Prior to a test an instrumentation plan should be formulated which includes the type of instrumentation, location, and accuracy. Typical accuracy tolerances of common measurements are as follows:

Speed	±0.5%
Torque	±2%
Power	±2%
Pressure	±2%
Temperature	±2%
Flow rate	±5%
Vibration frequency	±2%
Vibration amplitude (sine)	±10%

Instruments should be calibrated periodically and whenever possible an instrument reading should be verified by an independent measurement to determine if it is operating properly. For instance, if a flowmeter is used to measure oil flow, at some point a known volume of oil should be collected and the time recorded to check the flowmeter accuracy. Instrumentation failure is common and must be anticipated; therefore, redundant readings and backup systems must be planned for.

The most convenient way to measure speed is with a toothed wheel incorporated in the shaft system. An electromagnetic transducer is activated by the teeth passing by and the rpm are digitally displayed via an electronic counter. Care must be taken to set the proper gap between the tooth tips and the transducer. Also, runout of the gear can give erroneous readings. The counter gear usually has 60 teeth, but any number of teeth can be used if the electronics are set up to compensate. Other ways to measure speed are with a hand-held tachometer placed in contact with a shaft, and with strobe lights. When conducting a test, speed should always be measured in two different ways to ensure that a mistake is not made. At initial startup the correct direction of rotation should be verified.

Torque can be measured by monitoring shaft twist with strain gages. An indirect way of determining torque is by measuring the forces at the gearbox mounting pads (Figure 11.11). The housing torque is either the sum or difference of the input and output torques depending on whether the input and output shafts are rotating in the same or opposite directions. In Figure 11.11 assume the following:

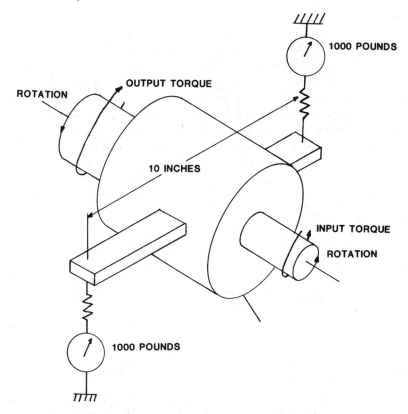

Figure 11.11 Gearbox housing torque measurement.

$$F = 1000 \, lb$$

$$D = 10 \, in.$$

therefore, the housing torque is 10,000 in.-lb. If the gear box ratio is 2:1, the input torque T_i is one-half the output torque T_o, and if the input and output shafts rotate in the same direction:

$$T_{HSG} = T_o - T_i = \frac{1}{2}T_o$$

and the output torque is twice the housing torque, or 20,000 in.-lb.

Temperature, pressure, and vibration measurements are covered later in the section "Condition Monitoring."

Reporting

The effort expended in conducting a test may be wasted if the test procedures and results are not carefully documented. The test items should be documented with part and serial numbers. Test setups should be described with photographs and sketches. Test results should be presented completely in the form of tables and graphs. If failures occur, they should be extensively documented as well as the modification incorporated to resolve the problem. The final report may be referred to years after the test program has been completed and the memory dimmed. Following is a general outline for a final report:

Objective
Summary
Test equipment
Test item
Test plan
Results
Discussion
Conclusions and recommendations

SPECIAL TESTS

Sound Testing

Measurement of the sound level of gearboxes is becoming increasingly important as government standards concerning equipment noise generation become more stringent. In addition to noise testing a brief discussion of sound fundamentals is presented in this section.

As a gearbox vibrates, a pressure oscillation in the surrounding medium (usually air) is generated. The transmission of the pressure vibration is called a sound wave and as it travels through the medium it can be detected by some form of receiver such as a microphone or the human ear. The sound pressure level of a gearbox is conventionally specified in decibels (dB) at a given distance from the gearbox. The sound pressure level L_p is the ratio of the pressure of the sound being measured to a reference pressure:

$$L_p = 20 \log_{10} \frac{p}{p_0}$$

where

p = sound pressure being measured, N/m^2
p_0 = reference pressure, N/m^2

(*Note*: psi $\times$ 6893 = N/m^2.)

The reference pressure is conventionally taken as 20 μN/m^2 (20 $\times$ 10^{-6} N/m^2), which is approximately the threshold of normal hearing at a frequency of 1000 Hz. As an example, if the sound pressure 2 ft from a gearbox is measured as 0.0010 psi (6.893 N/m^2), the sound pressure referred to 20 μN/m^2 is

A, B, AND C ELECTRICAL WEIGHTING NETWORKS FOR THE SOUND-LEVEL METER

These numbers assume a flat. diffuse-field response for the sound-level meter and microphone.

FREQUENCY Hz	A-WEIGHTING RELATIVE RESPONSE, dB	B-WEIGHTING RELATIVE RESPONSE, dB	C-WEIGHTING RELATIVE RESPONSE, dB	FREQUENCY Hz	A-WEIGHTING RELATIVE RESPONSE, dB	B-WEIGHTING RELATIVE RESPONSE, dB	C-WEIGHTING RELATIVE RESPONSE, dB
10	−70.4	−38.2	−14.3	500	−3.2	−0.3	0
12.5	−63.4	−33.2	−11.2	630	−1.9	−0.1	0
16	−56.7	−28.5	−8.5	800	−0.8	0	0
20	−50.5	−24.2	−6.2	1,000	0	0	0
25	−44.7	−20.4	−4.4	1,250	+0.6	0	0
31.5	−39.4	−17.1	−3.0	1,600	+1.0	0	−0.1
40	−34.6	−14.2	−2.0	2,000	+1.2	−0.1	−0.2
50	−30.2	−11.6	−1.3	2,500	+1.3	−0.2	−0.3
63	−26.2	−9.3	−0.8	3,150	+1.2	−0.4	−0.5
80	−22.5	−7.4	−0.5	4,000	+1.0	−0.7	−0.8
100	−19.1	−5.6	−0.3	5,000	+0.5	−1.2	−1.3
125	−16.1	−4.2	−0.2	6,300	−0.1	−1.9	−2.0
160	−13.4	−3.0	−0.1	8,000	−1.1	−2.9	−3.0
200	−10.9	−2.0	0	10,000	−2.5	−4.3	−4.4
250	−8.6	−1.3	0	12,500	−4.3	−6.1	−6.2
315	−6.6	−0.8	0	16,000	−6.6	−8.4	−8.5
400	−4.8	−0.5	0	20,000	−9.3	−11.1	−11.2

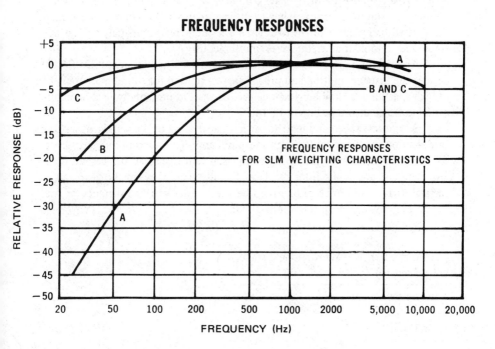

Figure 11.12 Sound meter frequency response. (From Ref. 1.)

$$L_p = 20 \log_{10} \frac{6.89 \text{ N/m}^2}{20 \times 10^{-6} \text{ N/m}^2} = 110.7 \text{ dB}$$

The sound-level meters that are used to measure decibels have a frequency response which is better than the human ear. In other words, at specific frequencies the meter will record sound a human being would not hear. To compensate for this fact, various filters have been incorporated into sound meters so that the measurements will approximate what the ear would record. As shown in Figure 11.12, three weighing scales, A, B, and C, have been established. The A scale matches the human ear's response at sound levels below 55 dB, the B scale at levels between 55 and 85 dB, and the C scale above 85 dB. The A scale is most commonly specified due to its use by OSHA for measurements up to 115 dB. Therefore, in the example above, the gearbox sound level measured by a meter set on the A scale would be defined as 110.7 dBA.

In order to analyze the source of gearbox noise, sound meter readings can be filtered to register only a limited range of frequencies. Frequently, octave and 1/3-octave bands are specified, as defined in Table 11.1.

Figure 11.13 presents the results of an octave band analysis of a two-stage parallel shaft gearbox with the high-speed mesh operating at 18,000 rpm. The lower curve is a measure of the background noise, including the prime mover decoupled from the gearbox. When even finer analysis of noise data is required, filters with bandwidths down to 2 Hz are used. A real-time analyzer can be employed to look at all frequencies simultaneously.

American Gear Manufacturers Association (AGMA) Standard 295.04 [2] defines the instrumentation and procedures to be used for sound measurement of high-speed helical and herringbone gear drives and presents typical maximum sound levels as shown in Figure 11.14.

Sound test speed and load conditions, according to the standard, are to be agreed upon by the manufacturer and purchaser. The microphone is to be located perpendicular to the center of the gear unit's vertical surface, but not less than 1 ft above the floor or plate. The distance between the gear unit's vertical surface and microphone is to be the normal working distance of the closest employee or as in the following table:

Gearbox center distance (in.)	Microphone distance (ft)
4 and below	3
4-15	5
15 and over	10

Table 11.1 Continuous Octave and One-Third Octave Frequency Bands

	Frequency (Hz)					
	Octave			One-third octave		
Band	Lower band limit	Center	Upper band limit	Lower band limit	Center	Upper band limit
12	11	16	22	14.1	16	17.8
13				17.8	20	22.4
14				22.4	25	28.2
15	22	31.5	44	28.2	31.5	35.5
16				35.5	40	44.7
17				44.7	50	56.2
18	44	63	88	56.2	63	70.8
19				70.8	80	89.1
20				89.1	100	112
21	88	125	177	112	125	141
22				141	160	178
23				178	200	224
24	177	250	355	224	250	282
25				282	315	355
26				355	400	447
27	355	500	710	447	500	562
28				562	630	708
29				708	800	891
30	710	1,000	1,420	891	1,000	1,122
31				1,122	1,250	1,413
32				1,413	1,600	1,778
33	1,420	2,000	2,840	1,778	2,000	2,239
34				2,239	2,500	2,818
35				2,818	3,150	3,548
36	2,840	4,000	5,680	3,548	4,000	4,467
37				4,467	5,000	5,623
38				5,623	6,300	7,079
39	5,680	8,000	11,360	7,079	8,000	8,913
40				8,913	10,000	11,220
41				11,220	12,500	14,130
42	11,360	16,000	22,720	14,130	16,000	17,780
43				17,780	20,000	22,390

Source: Ref. 1.

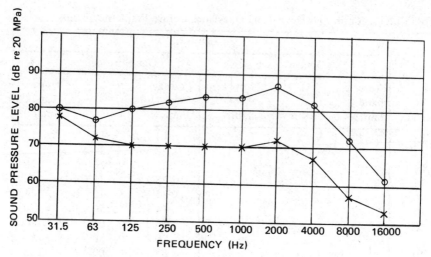

Figure 11.13 Octave band analysis (end location at 3 ft and 14,200 rpm).

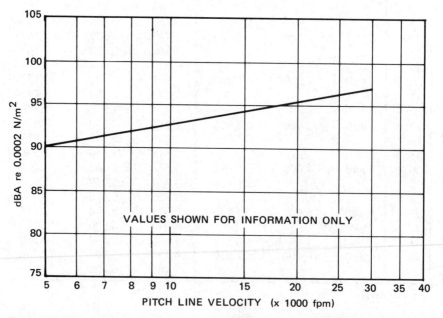

Figure 11.14 Typical maximum sound pressure levels versus high-speed mesh pitch line velocity. Note: In case of multireduction or increasing gear sets within one housing use the pitch line velocity of the highest speed set. (From Ref. 2.)

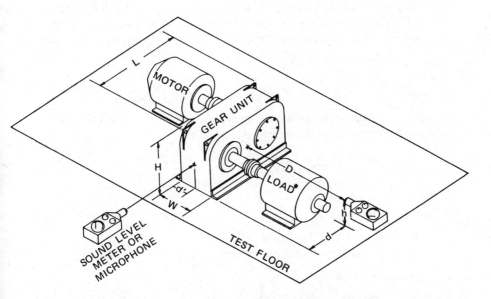

Figure 11.15 Sound test microphone position. Key: L, length of gear unit; H, height of gear unit; W, width of gear unit; D, distance of microphone perpendicular to unit, as specified in standard for size, h, height of microphone perpendicular to floor (H/2); d, distance of microphone from corner of unit (1/2) or (W/2). *Note: Load is optional for factory testing. (From Ref. 3.).

Figure 11.15 shows the sound test microphone position. The sound level is to measured with and without the gear unit operating so as to correct the unit rating for the ambient sound pressure level. Corrections are presented in Table 11.2.

Table 11.2 Corrections for Ambient Sound Pressure Levels

Difference between gear unit and ambient sound pressure levels (dB re 20 μN/m^2)	Correction to be subtracted from gear unit sound pressure level (dB re 20 μN/m^2)
3 or less	3
4 and 5	2
6–9	1
10 or greater	0

Source: Ref. 2.

The overall sound level of a mechanical system is made up of the sound levels of the components therein. For instance, the components of a generator drive system when measured individually may exhibit the following noise levels:

Prime mover: 88 dBA
Gearbox: 82 dBA
Generator: 95 dBA

The following expression is used to combine the component noise levels to analytically define an overall noise level:

$$L_p = 10 \log_{10} \sum_{i=1}^{N} 10^{0.1 a_i}$$

where

N = number of components
a_i = sound level of each component

In the example,

$$L_p = 10 \log_{10} (10^{8.8} + 10^{8.2} + 10^{9.5})$$
$$= 96 \text{ dBA}$$

It can be seen that in order to reduce the noise level of this system, the major contributor, the generator, must be worked on.

Noise generation is influenced by the design and manufacture of the gearbox components. The ideal situation in a gear mesh would be to transmit power with no change in the angular velocity of the gear shaft. In such a perfect gear mesh there would be no accelerations or decelerations of the gear shafts, which provide the energy for vibration and noise generation. Of course, there are no perfect gear meshes, and errors in tooth spacing, profile, or runout will always be present, resulting in accelerations and decelerations of the gear shafts and noise.

Design of gear teeth can reduce noise levels. For instance, increasing overlap or changing from spur to helical gearing will reduce noise. In these cases the changes result in smoother load distribution from tooth to tooth as they go through the mesh. Changes in design to reduce noise, however, might adversely affect other parameters. For instance, increasing contact ratio by going to a finer pitch might compromise bending strength. Other design changes that tend to reduce noise are:

Reduced pitch line velocity
Proper profile and lead modifications
Lower pressure angles

It must be realized that detail gear tooth geometry changes can only account for changes in sound level of up to approximately 4 dBA. Also, no matter how optimum the design, the quality of the gearing will determine the sound level. Table 11.3 lists the common sources of sound in a gear drive system.

Sound can be transmitted through the air or be structure borne. Structure-borne noise may travel through the support structures and radiate at some point other than its source. Because of this the noise measurements near a gearbox in an operating system may be quite different from those measured during a bench test. Also, when connected to the prime mover and load, the torsional response of the system may be different from the bench test and result in different noise measurements.

In many cases when it is determined that the noise generation of a system is excessive yet not an indication of some malfunction, the most practical method of noise reduction is the use of an acoustical enclosure. The effectiveness of such an enclosure is very dependent on eliminating all openings. Such an enclosure will also affect the ambient temperature and housing heat dissipation ability of the gearbox and means must be provided to cool the unit properly.

Efficiency Testing

Gearboxes are an extremely efficient means of transmitting power. Depending on the operating conditions, design, and manufacturing techniques, the loss per mesh of spur, helical, or bevel gears will vary between approximately ½ to 3%. Analytical means are available to calculate gearbox efficiency, but in order to be confident that an accurate measure of a unit's efficiency is known, the value must be arrived at by testing.

Obviously, the efficiency of the gearbox affects the total system efficiency and the rating of the prime mover required. The power lost in the gearbox, however, also has a large impact on system design since it must be dissipated as heat.

Many low-speed gearboxes have a self-contained splash lubrication system and the power rating of the unit may be limited by the oil temperature rise rather than a mechanical limit. This is the reason gearbox catalogs list a thermal rating in addition to a mechanical rating. When the heat generated in a unit is greater than can be dissipated through the casing an external lubrication system including a cooler is required. The heat generation or efficiency of the gearbox will determine the size of cooler required.

The heat generation can be calculated as follows:

$$Q = MC_p \, \Delta T$$

where

Q = heat generated, Btu/min
C_p = specific heat, Btu/lb-°F

Table 11.3 Common Sources of Airborne and Structure-borne Sounds Generated in Gear Drive Systems[a]

Instruments that provide the operator with not only the amplitude of the vibration or noise, but also the predominant frequencies can be a tremendous aid in determining sources. These causes normally present themselves as follows:

1. *Balance:* Residual unbalance presents itself at a frequency equal to once per shaft revolution and it will increase in amplitude as speed is increased.

2. *Alignment:* Misalignment will present itself at once or some-times twice and three times per shaft revolution. However, the amplitude will remain fairly constant with speed changes.

3. *Friction:* This is difficult to pinpoint by vibration and noise frequency. Amplitude may be very high when continuous sliding occurs. It may also be random, high-amplitude, shock-type pulses, as in hydrodynamic bearing rubbing. It may be irregular and often violent.

4. *Looseness:* This may cause unbalance, misalignment and friction rubbing at moderate and high speeds. At low speeds, it may display itself as an irregular rattle. Often it shows up at twice shaft rotational speed.

5. *Distortion:* This is often an indirect cause of vibration and noise, which also leads to unbalance, misalignment, or friction. It will tend to change in amplitude with load or operating temperatures, when speed is held constant.

6. *Critical speeds:* These occur through any given speed range and are points at which a rotating system likes to vibrate torsionally or laterally at a particular frequency. Rotors characteristically show violent increase in amplitude at particular critical speeds but are fairly stable above and below these speeds. A critical speed may change frequency with load and temperatures.

7. *Resonances:* These also display themselves as frequencies at which system members like to vibrate. The distinction from critical speeds is that resonances occur in other than rotating members, and affect alignment. Resonances occur at fixed frequencies and change in amplitudes with load and temperature.

8. *Tooth mesh (i.e., tooth contact):* This will show up at tooth mesh frequency (i.e., rotating speed times number of teeth) and multiples of this mesh frequency.

9. *Bearing instability:* Bad antifriction bearings will cause high-frequency vibration at several times rotational speed, also friction vibration will occur. Hydrodynamic bearings, lightly loaded, will tend to whirl at 0.43 to 0.47 times the rotational speed. This so-called "half-frequency whirl" will "on-set" violently with speed or temperature changes and may continue until the rotor is completely stopped.

10. *System pulses:* These may occur in many types of systems, such as the vane-pass frequency of a pump or compressor (rotational speed times the number of vanes), and the beating of reciprocating engines which cause frequencies at one-half and one-quarter rotational speed at various amplitudes.

11. *Windage:* Couplings and other rotating parts generally create broadband noise, but can be at a bolt pass frequency or fan blade pass frequency.

[a]All of these types of vibrations and noise frequencies can be generated in a gear drive. Major frequencies can interact and cause frequency modulation and phase shifts. Any combination, sum difference, and multiple (harmonics) of prime frequencies can occur if the forcing magnitude and system freedoms are such that they will cause and allow the generated vibration to become predominant. Generally, only the prime frequencies will present themselves as problem modes. However, sometimes very elusive frequencies appear, such as periodic cutting machine error appearing on one of the gears.
Source: Ref. 3.

M = oil flow, lb/min

ΔT = oil temperature rise across the gearbox, °F

Note:

$$\frac{Btu/min}{42.44} = horsepower$$

C_p of oil $\cong$ 0.5 Btu/lb-°F

1 gpm $\cong$ 7.5 lb/min of oil

By accurately measuring the temperature of the oil entering and leaving the gearbox and the oil flow, the heat rejected to the oil can be calculated as shown above.

For example, a gearbox with oil in temperature of 130°F, oil-out temperature of 160°F and oil flow of 20 gpm will reject 848 Btu/min (20 hp) to the oil. If the gearbox is transmitting 1000 hp, the efficiency can be calculated as follows:

$$E = \frac{P_t - P_1}{P_t} = \frac{1000 - 20}{1000} = 98\%$$

where

P_t = total hp transmitted

P_1 = power loss

It must be realized that in addition to the heat rejected to the oil, heat is dissipated to the atmosphere by radiation and convection through the casing. To arrive at an estimate of the heat dissipated through the casing, the following heat transfer equation may be used:

$$H = CA(Tc - Ta)$$

where

H = heat dissipated by the gear casing, Btu/hr

C = combined coefficient of radiation and convection, Btu/hr/ft/°F

Tc = casing temperature, °F

Ta = ambient temperature, °F

The value of C depends on the material, roughness, and color of the housing and the velocity of air around the housing and can vary from approximately 0.5 to 3.0. To eliminate the variable of casing radiation and convection, the gear housing can be insulated for an efficiency test. In this way all the heat generated in the unit will be carried away by the oil.

The heat balance is the most widely used method of measuring gear unit efficiency. A more direct way is to measure input and output torque, but this method requires exact and expensive instrumentation. In the case of a regenerative or back-to-back test, only input torque need be measured and this will monitor the torque required to drive the two locked-up gear trains. Since the prime mover supplies only enough power to make up the losses in the system, the torque lost in each gearbox is one-half the input torque.

Power losses in the gearbox can be divided in two categories, friction losses in the gear mesh and at the bearing interfaces and windage or churning losses. In high-speed boxes, windage and churning losses can be considerable amounting to up to half the total power loss. An easy way to determine these losses is to operate the gearbox at full speed, no load, and then perform a heat balance.

A high-speed box is considered to be one with pitch line velocity ranging from 5000 to 20,000 fpm. Above 20,000 fpm windage and churning can be a limiting factor to the success of the design. For instance, some wide-face-width helical gearboxes operating at pitch line velocities above 20,000 fpm have experienced severe heat generation problems due to pumping of oil and air along the face width. In high-speed double helical gears the hands of helix should always be selected so as to pump oil away from the apex.

When it is determined that a gearbox has excessive windage and churning losses, the situation can be relieved by strategically placing baffles or screens in the unit. In low-speed units using splash lubrication it is beneficial to shroud the gears dipping into the sump oil. With high-speed units the sump oil level should be below the gears and the sump should be separated from the gear meshes by a baffle plate or screen. Minimizing windage and churning losses is more an art than a science and can require trial-and-error solutions verified only by testing.

Lubrication system parameters can have a profound effect on gearbox efficiency. For instance, efficiency will increase with reduced oil flow; obviously, this is due to reduced churning. The limit to this technique is the increase in temperature rise across the box and in the individual components as oil flow is reduced. Increasing oil in temperature will increase efficiency since the oil viscosity will decrease and therefore reduce churning. Changing to a lower-viscosity oil will have the same effect. Proper scavenging of the unit is extremely important in minimizing churning; therefore, great care must be taken in the location and size of drain holes. Also, back pressure must be minimized.

INITIAL FIELD STARTUP

Prior to starting the equipment, the following preliminary checks should be performed:

1. Check oil level and ensure that the proper oil is being used.
2. Tighten all pipe connections.
3. Check all electrical connections.
4. Tighten all mounting and gearbox bolts with proper torque.
5. Check mounting of all gauges, switches, and so on.
6. Check all couplings for proper installation and alignment.
7. Check inspection cover installations.

The following instructions pertain to the initial startup:

1. The unit should be preoiled to ensure lubrication of the journal bearings at startup.
2. The gearbox should be started slowly under as light a load as possible. Observe that the rotation is in the proper direction. Check the system oil pressure.
3. After starting, when the oil has been circulated, the unit should be stopped and sufficient oil added to bring the sight gage oil level up to the specified amount.
4. As the unit is brought up to operating speed, it should be continuously monitored for excessive noise, vibration, or temperature. If any of these occur, shut down immediately, determine the cause, and take corrective action. Also check for oil leaks as the unit is initially operated.
5. If possible, operate at half load for the first 10 hr to allow final breaking in of the gear tooth surfaces.
6. After the initial 50 hr of operation the oil in a new unit should be drained and the case flushed with SAE 10 straight mineral flushing oil containing no additives. Drain the flushing oil and refill with the recommended lubricant to proper level.
7. After the initial 50 hr of operation check all coupling alignments and retorque all bolts. Check all piping connections and tighten if necessary.

If starts are made in a cold environment, consideration should be given to preheating the lubricant. Load should not be applied until the lubricant has attained operating temperature.

CONDITION MONITORING

On an operational gearbox the question always arises as to how to evaluate the condition of the unit and when to disassemble the equipment and inspect or replace internal components. Figure 11.16 presents a hazard function or what is traditionally called the "bathtub curve," which describes the failure rate of a unit at any particular point in its operating history. Initially, there is a high "infant mortality" period, which reflects failure due to assembly or

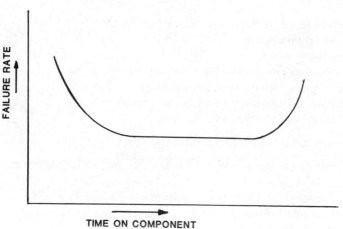

Figure 11.16 Hazard function.

manufacturing errors. For instance, an oil jet might have been clogged or a bearing preloaded at assembly. These early failures can often be screened out during an acceptance test program. Following the "infant mortality" phase, there is a period of constant failure rate where failure modes related to time appear. Examples of these are bearing or gear tooth fatigue, spline wear, and so on. At first glance it would appear reasonable that the unit be removed from service at the point in time where the failure rate begins to increase and the components inspected and reworked or replaced as necessary. This is not necessarily so, since removal and disassembly of the unit begins another operating cycle and the gearbox again would be subject to the high "infant mortality" failure rate. As a practical matter it is difficult to establish the point where the failure rate increases sufficiently to warrant overhaul since this requires an extensive data base. Gear units are subject to many different modes and within any given failure mode there will be significant scatter.

The most efficient method of determining when a gear unit requires service is to base repairs on the condition of the gearbox rather than overhaul it at an arbitrary time period. The basic idea is not to disturb equipment that is operating properly but inspect only items that exhibit potential failure symptoms. This type of predictive maintenance is termed "on condition" and requires instrumentation of a gear unit and the proper interpretation of the data provided. The goals of this technique are threefold:

1. Detect gear units which are operating abnormally.
2. Diagnose which internal component is deteriorating.
3. Predict how long the unit can function in this condition before corrective action must be taken.

Condition monitoring systems are expensive and require discipline to set up and use effectively. It is difficult to define the normal operating parameters of a system and this initial step is vital in an "on condition" program. Sometimes the normal operating parameters are termed the "signature" of a component. Once this is established it is relatively easy to identify abnormal operation. When an "on condition" program is achieved, the benefits are as follows:

1. Reduction in downtime since potentially catastrophic failures are detected and corrected. This also leads to a reduction in repair costs.
2. Enables efficient scheduling of equipment shutdown since warning of a failure is detected well in advance of the event.
3. Shutdown periods are shorter because unnecessary work is avoided and the work required is known in advance so that preparations can be made.

In the following pages the various methods used to monitor gearbox condition will be discussed. These include:

Vibration
Noise
Shaft position
Oil temperature
Oil pressure
Oil analysis
Chip collection

Oil Analysis

Gearbox condition can be monitored by the analysis of oil samples. One widely used technique is SOAP (Spectrographic Oil Analysis Program). An oil sample is taken periodically from the gearbox sump and sent to an analytical laboratory, where it is burned and the light waves passed through a spectrometer. The spectrum of light waves given off by the sample yields information as to the types of wear metals suspended in the oil and their quantity.

In principle, if it is possible to identify and measure the wear metals present in the lubricant, one can identify which internal gearbox component is deteriorating. Also, by monitoring the rate of change of wear metal from sample to sample, conclusions can be drawn concerning the gearbox condition and when maintenance will be required. Figure 11.17 illustrates the information presented in a typical SOAP laboratory analysis.

To use SOAP properly, a gearbox history base must be established to determine what type and quantity of wear metals the unit normally generates. Also, a listing of materials used in the gearbox must be available. With this information SOAP can be used to predict failures. For instance, an increase in silver might mean that a silver-plated antifriction bearing retainer is wearing. Tin and

DATE		HOURS		OIL ADDED	PHYSICAL PROPERTY TESTS			SPECTROCHEMICAL ANALYSIS METALS:PARTS PER MILLION BY WEIGHT																	
SAMPLED	RECEIVED	OIL	UNIT		VIS. AT 100°F	TOTAL ACID NO.	WATER	IRON	LEAD	COPPER	CHROMIUM	ALUMINUM	NICKEL	SILVER	TIN	SILICON	BORON	PHOSPHORUS	SODIUM	ZINC	CALCIUM	BARIUM	MAGNESIUM	TITANIUM
1/25	1/25	48.3	48.3	—	29.2	0.0	<.05	.02	.01	.0	.02	.0	.0	.0	.0	.0	.0	63.	.0	20.	.0	.0	.0	.0
2/1	2/10	76.3	76.3	—	28.9	0.03	<.05	.04	.01	.0	.02	.0	1.0	.0	.0	11.	1.	81.	7.	20.	50.	20.	.0	.0

Figure 11.17 Typical oil analysis form.

lead traces could indicate a journal bearing problem. A large percentage of iron might point to gear or bearing distress. Of course, practically all gearbox components contain some iron, but if other elements are also indicated, the cause of distress might be narrowed down to a specific component.

SOAP oil samples should always be taken from the same area of the unit. Oil sampling periods are in the order of every 10 to 100 hr of operating time and each oil sample should be placed in a clean vial marked with an identifying number, the date, and the number of operating hours of the unit. If readings indicate a problem, the sampling interval should be reduced. There is a time delay in getting results from the laboratory and a typical turnaround time might be 3 days.

SOAP readings can be influenced by external factors. For instance, oil changes or oil additions must be accurately recorded since they will reduce the percentage of wear metals. Also, the degree of filtration will affect the wear metals suspended in the fluid. Very fine filtration such as 3 μm will effectively remove all wear metals and make SOAP meaningless. It should also be recognized that large particles will settle out and not be detected by SOAP analysis. The effective detection range of the SOAP procedure is with particles in the size range 1 to 10 μm.

Increases in the wear metal percentage between samplings are more significant than the total wear metal content since an increase signifies a trend of abnormal operation. Therefore, a rapid increase in wear metal should be investigated even if the absolute percentage does not exceed the acceptance criterion.

When performing a SOAP analysis other oil characteristics are commonly monitored, such as acid number, viscosity, and water content. The acid number is relatively easy to check and will increase due to the presence of wear metals or due to overheating. Occasionally, one can detect acid oil by the odor alone, which is distinctive and unpleasant. Changes in oil viscosity indicate lubricant deterioration and can be monitored. Monitoring of acid number and/or viscosity can be done more quickly and economically than SOAP but does not yield as much information.

Vibration

Gear boxes are mass elastic systems and therefore will vibrate when excited by internal or external influences. Vibration of gear units can be measured in several different ways. The motion of the shafts, both radial and axial, can be observed with noncontacting sensors. Casing motions can be measured using velocity or acceleration pickups. Generally, the following parameters are monitored:

1. Peak-to-peak mils displacement
2. Peak inches per second velocity

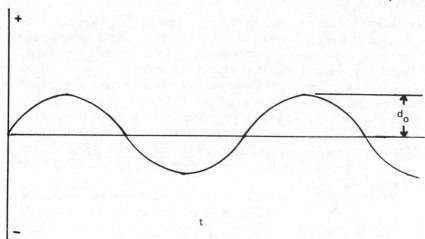

Figure 11.18 Simple harmonic motion.

3. Peak g's acceleration
4. Frequency of vibration

The relationship of vibratory displacement, velocity, and acceleration is as follows:

A vibration of simple harmonic motion which has a pure sinusoidal waveform has a displacement d (see Figure 11.18):

$$d = d_0 \sin \omega t$$

where

d_0 = ½ peak-to-peak displacement reading, in.
ω = frequency, rad/sec
    = (2π)rpm/60
t = time, sec

Differentiating for the velocity v:

$$v = \omega d_0 \cos \omega t$$

and differentiating once more for the acceleration a:

$$a = -\omega^2 d_0 \sin \omega t$$

To calculate the G loading, let $\sin \omega t = 1$, and divide a by the acceleration of gravity:

$$G = \frac{\omega^2 d_0}{g}$$

where g is 386 in./sec^2.

The major contributors to gearbox vibration are:

1. Rotating shaft unbalance
2. Shaft assembly problems such as loose connections, bent shafts, and misalignment
3. Gear tooth inaccuracies
4. Component wear

AGMA Standard 426.01 [4] presents acceptable vibration levels, as shown in Figure 11.19. If the displacement amplitude is not obtainable at discrete frequencies, the standard allows either of the following:

1. A nominal unfiltered velocity level of 0.3 in./sec but not exceeding a maximum 2 mil displacement. (*Note*: a mil is 0.001 in.)
2. An unfiltered displacement level determined from Figure 11.19 using the shaft rotation speeds as discrete frequencies (i.e., 120 rpm = 20 Hz).

A gearbox may be acceptable according to the criteria of Figure 11.19 in acceptance testing at the manufacturer's facility, yet exhibit higher vibration levels in field service. The vibration levels may be adversely affected by factors not under control of the gear manufacturer, such as:

1. Inadequate foundation
2. Excessive shaft misalignment
3. Coupling components not tested with the gear unit
4. Resonance of base or other supporting structure
5. System torsional vibration
6. Motor magnetic center wobble
7. Unbalance or other forced vibration from other components in the system

Shaft vibration levels as shown in Figure 11.19 are measured by the use of non-contacting, eddy current proximity devices. A probe is positioned approximately 0.050 in. from a shaft and the distance (gap) between the probe tip and the shaft is translated to a voltage.

A proximity probe is basically a small coil of wire embedded in a ceramic tip on the end of a holder. This coil is supplied with a high-frequency voltage which produces a magnetic field radiating from the probe tip. Any conducting surface that lies within this magnetic field absorbs some of its power, and the closer to the tip the more power absorbed. The RF signal to the coil is supplied by a proximitor which has the secondary function of relaying the resultant field strength back to the readout instruments by reconverting it to a voltage.

Proximitors and probes are calibrated to give a linear relationship between readout voltage and distance of observed material. The range of this linear relationship depends on the type of probe and supply voltage, but is on the order of

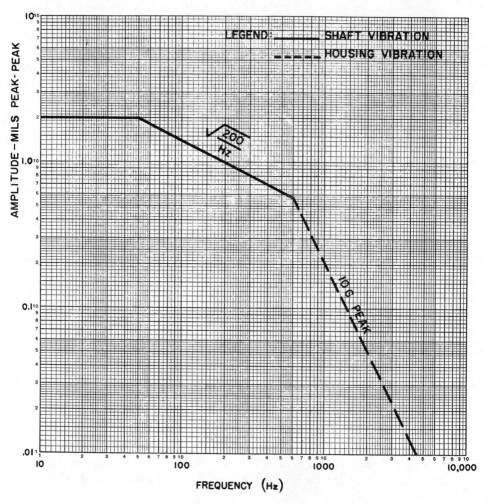

Figure 11.19 Acceptable shaft vibration levels. (From Ref. 4.)

80 mils. The readout obtained from a radially mounted probe is the electrically measured difference between the minimum and maximum distances of the probe from the shaft material (peak-to-peak amplitude).

The American Petroleum Institute has issued a standard [5] which defines the use of proximity probes. Figures 11.20 to 11.22 illustrate radial and axial probe arrangements and their application in a gearbox.

Note that in Figure 11.22 two radial probes 90° apart are mounted at the bearing positions. It is possible to have totally different vibration in two

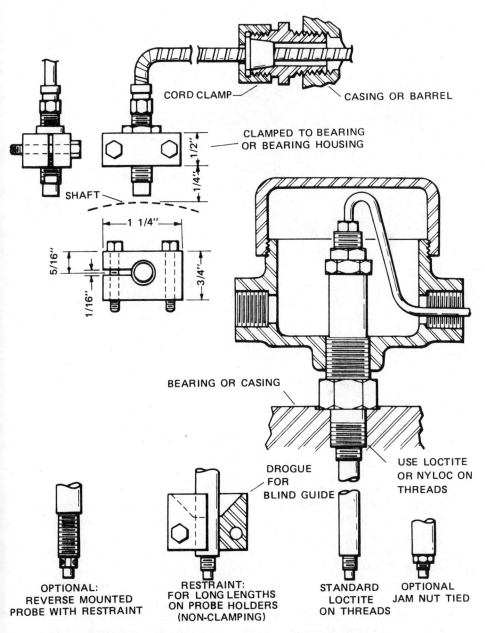

CORD CLAMP

CASING OR BARREL

CLAMPED TO BEARING
OR BEARING HOUSING

1/2"

1/4"

SHAFT

1 1/4"

5/16"

3/4"

1/16"

BEARING OR CASING

DROGUE
FOR
BLIND GUIDE

USE LOCTITE
OR NYLOC ON
THREADS

OPTIONAL:
REVERSE MOUNTED
PROBE WITH RESTRAINT

RESTRAINT:
FOR LONG LENGTHS
ON PROBE HOLDERS
(NON-CLAMPING)

STANDARD
LOCTITE
ON THREADS

OPTIONAL
JAM NUT TIED

Figure 11.20 Typical radial probe arrangement. (From Ref. 5.)

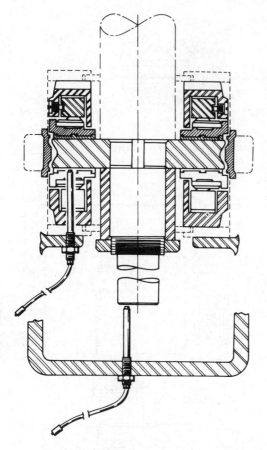

Figure 11.21 Typical axial probe installation. (From Ref. 5.)

perpendicular directions at one particular bearing. This is generally true of any
vibration measurement, whether it be at the bearing or at the casing and readings
in two planes should always be taken. Gearbox bearing probes are mounted 45°
from the vertical center since quite often there is a split line at the horizontal.

On Figure 11.22 a phase angle probe is shown on each shaft. This is a
transducer that observes a once-per-turn event such as a keyway. Its function is
to provide a reference mark and timer for speed, phase angle, frequency
measurements, and all data acquistiion. Figure 11.23 shows a phase angle probe
installation. Each time the keyway passes the probe a voltage pulse results. This
pulse provides a physical reference on the shaft which can be used to measure
the high spot of the shaft.

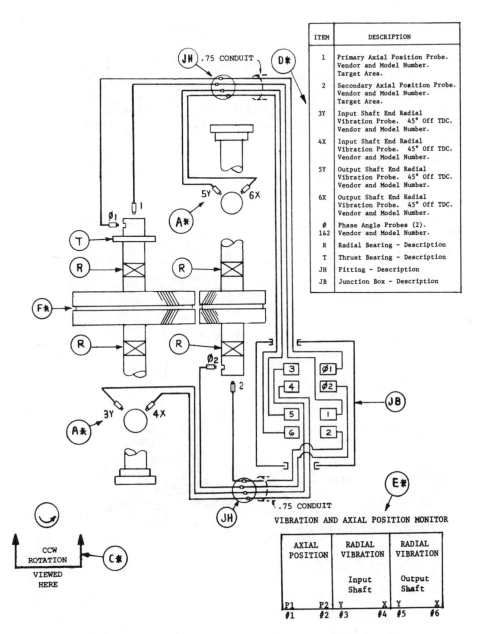

Figure 11.22 Typical system arrangement for double helical gear. (From Ref. 5.)

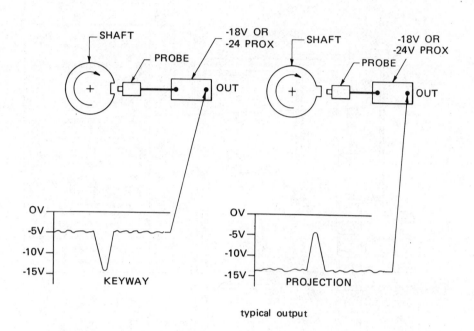

typical output

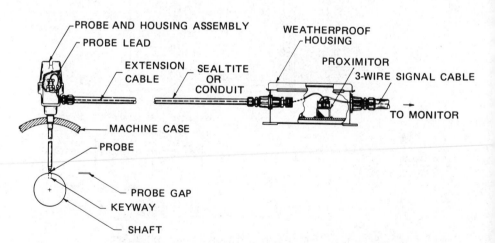

Figure 11.23   Phase angle probe. (Courtesy of Bently Nevada, Minden, Nevada.)

To illustrate the type of data acquired by proximity probes, refer to Figure 11.24. The oscilloscope photograph displays unfiltered time-domain traces from both horizontal and vertical probes, as well as the same signals filtered at rotational frequency. The inclusion of a phase angle mark, K∅ , identifies one complete revolution of the shaft. In the middle of the picture the shaft orbit can be observed as an unfiltered motion, and filtered exactly at running speed. In addition, a spectrum analysis of each signal is presented, and the various components identified.

There are two types of oscilloscope readings presented in the center of Figure 11.24. One is the time-base mode, where the sinusoidal-type waveform representing the shaft motion is displayed. This shows the position of the shaft relative to the input transducer versus time horizontally across the cathode ray tube of the oscilloscope. The other display is the orbit presentation where the output from two separate proximity probes at 90° to one another are shown in the X-Y mode of the oscilloscope. In this mode the centerline motion of the shaft is displayed. If the probes are mounted at the bearing, the orbit is a presentation of the motion of the shaft centerline with relationship to the bearing.

When measuring shaft vibration with proximity probes it is possible to get erroneous readings caused by physical and mechanical deformities in the shaft material. These erroneous readings are sometimes referred to as glitches.

Obviously, there will always be some mechanical runout in the shaft. This may be measured by a dial indicator mounted in the probe area of the shaft, which itself is supported at the bearing journals in V blocks or rollers. The shaft runout at the probe locations may be subtracted from the vibration readings to arrive at actual vibration levels, provided that the runout is shown to be in phase with the vibration.

In addition to runout, other mechanical problems that will cause glitches are bowed shafts and surface imperfections such as scratches, dents, and burrs. To correct these problems the shaft surface may have to be remachined or reground. If the surface irregularities are very minor, a redressing of the surface with an oil-wetted fine emery cloth may eliminate the surface imperfections.

In addition to mechanical runout there is the possibility of electrically induced runout due to residual magnetism which can be corrected by having the shaft degaussed. Other potential causes of glitch could be metallurgical segregation or residual stress concentrations.

Before assmbling a shaft that has probe areas a bench test should be conducted to ensure that glitch has been held to minimum (in the order of 0.00025 in.) so as not to influence vibration readings excessively. After functional inspection it is advisable to coat the probe areas with an epoxy resin which can remain in position for the life of the machine. This coating will not affect probe readings but will protect the probe area from corrosion and minor mechanical damage. The shaft position probes provide a vibration measurement of the

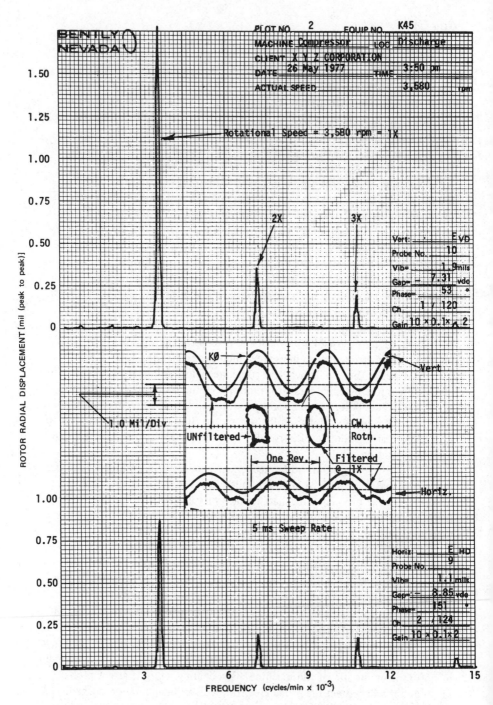

Figure 11.24 Shaft displacement data. (Courtesy of Bently Nevada, Minden, Nevada).

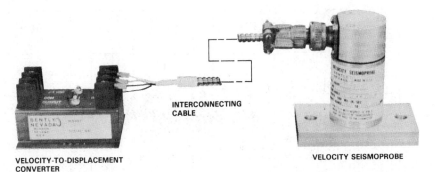

INTERCONNECTING CABLE

VELOCITY-TO-DISPLACEMENT CONVERTER

VELOCITY SEISMOPROBE

Figure 11.25 Velocity transducer system. (Courtesy of Bently Nevada, Minden, Nev.)

relative motion between the shaft and the mounting of the proximity probe. The probe is usually mounted rigidly to a bearing cap or the gear casing. In order to establish motion of the casing itself, velocity pickups are attached to the casing.

Figure 11.25 illustrates a velocity transducer system. The purpose is to measure gearbox casing or structural vibration velocity and convert the vibration velocity into an electrical signal that represents the displacement of the casing.

One type of velocity transducer works on the principle that as a magnet moves with respect to a relatively stationary coil, a current is induced in the coil. The magnet is rigidly attached to the pickup case and therefore vibrates along with the casing. The coil is suspended by sensitive springs inside the pickup case.

Figure 11.26 depicts the casing velocity characteristics emitted by a three-stage speed-increasing gearbox. The time-domain signal inset in the figure is quite complicated due to the transducer responding to a large number of excitations. A spectrum analysis of the signal reveals Fourier components that can be related back to the operational characteristics of the machine. Peaks can be seen at the rotational speeds of each gear plus the second harmonics of each fundamental frequency. As a rule of thumb, peak vibration velocity readings over approximately 0.5 in./sec signify extremely rough operation and warrant shutdown of the machinery. Readings below 0.1 in./sec indicate smooth, well-balanced, and well-aligned equipment. A reading of over 0.3 in./sec might be a good level at which to consider taking corrective action.

Casing vibration should be measured in the vertical, horizontal, and axial directions. The measurements should be taken on a rigid section on the housing. It should be noted that a measurement taken on a rigid casing can identify such things as structural or piping resonance, loose or cracked foundations, or external vibration input sources but may not transfer vibration amplitudes due

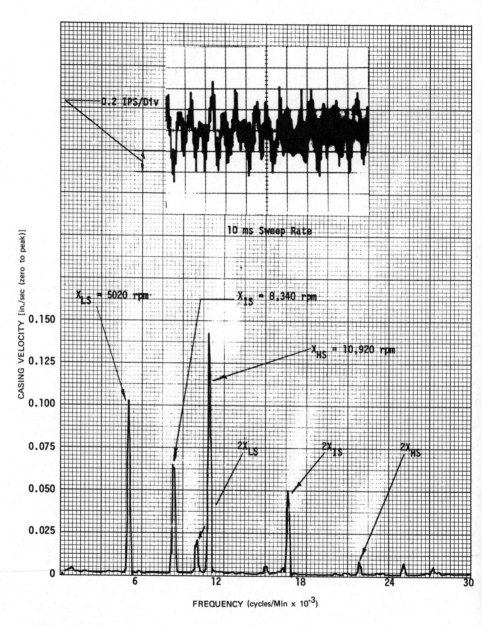

Figure 11.26   Gear box casing velocity characteristics. From Ref. 6; courtesy of Bently Nevada, Minden, Nev.)

Table 11.4 Vibration Conversion Factors

To obtain	Multiply the following by the numerical values:			
	Average	Rms	Peak	Peak to peak
Average	1.0	0.900	0.636	0.318
Rms	1.111	1.0	0.707	0.354
Peak	1.571	1.414	1.0	0.500
Peak to peak	3.142	2.828	2.0	1.0

to shaft motions. The casing may be too stiff to move as a result of shaft motion; therefore, on critical applications shaft position sensors should be incorporated in addition to casing pickups. When a casing transducer is located in the same plane as a shaft proximity probe, a vector summation of the two outputs can give the absolute shaft motion in addition to the shaft motion relative to the casing.

On occasion, rather than defining vibration in peak values, rms or average values are required. Table 11.4 presents the conversion factors. For high-frequency excitations such as gear meshing frequencies, accelerometers are used as transducers. Figure 11.27 shows an acceleration transducer system. The accelerometer portion of the acceleration transducer system is a "contacting" transducer that is physically attached to the vibrating machine part. The accelerometer uses a piezoelectric crystal situated between the accelerometer

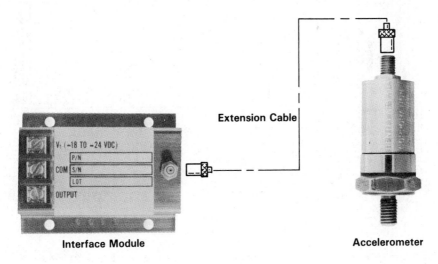

Interface Module   **Accelerometer**

Figure 11.27 Acceleration transducer system. (Courtesy of Bently Nevada, Minden, Nev.)

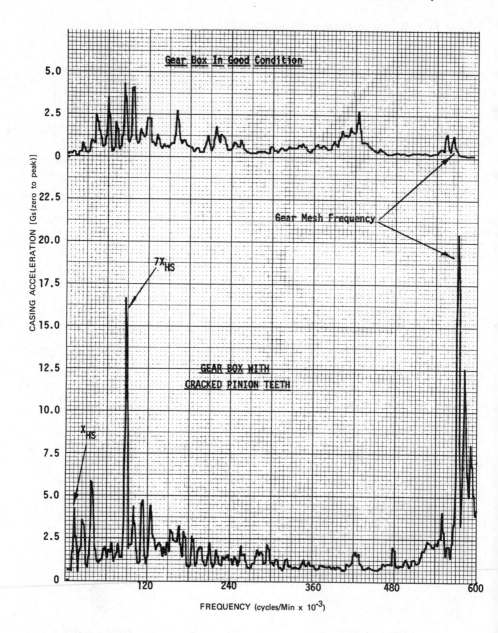

Figure 11.28 Gear box casing acceleration characteristics. (From Ref. 6; courtesy of Bently Nevada, Minden, Nev.)

base and an inertial reference mass. When the crystal is strained (compression or tension force), a displaced electric charge is accumulated on the opposing major surfaces of the crystal. The crystal element performs a dual function. It acts as a precision spring to oppose the compression or tension force and it supplies an electric signal proportional to the applied force.

Frequencies up to 30 kHz can be handled by accelerometers. Figure 11.28 illustrates gearbox casing acceleration characteristics. The top spectrum plot depicts a gearbox in good mechanical condition with reasonably low acceleration levels and a normal mixture of components. A similar measurement made on a unit that had cracked pinion teeth is presented on the bottom plot. High G loadings are exhibited at the gear mesh frequency, the seventh harmonic of pinion rotational speed, and the pinion running speed, X_{HS}.

Figure 11.29 Magnetic chip detector. (Courtesy of Technical Development Company, Glenolden, Pa.)

Figure 11.30 Chip collector with wear debris. (Courtesy of Technical Development Company, Glenolden, Pa.)

Chip Collection

A typical magnetic chip collector is shown in Figure 11.29. Figure 11.30 is a photograph of a chip collector with typical wear debris. The chip collector is installed in the oil sump or a scavenge return line. Figure 11.31 shows typical installations. A self-sealing valve allows withdrawal of the magnetic probe and visual inspection of the collected debris with loss of only a few drops of oil.

A system can be initiated where the chip collector is periodically inspected, every 25 to 50 hr, and records kept characterizing the particles collected. Once a baseline is defined, any change in the debris collected during a

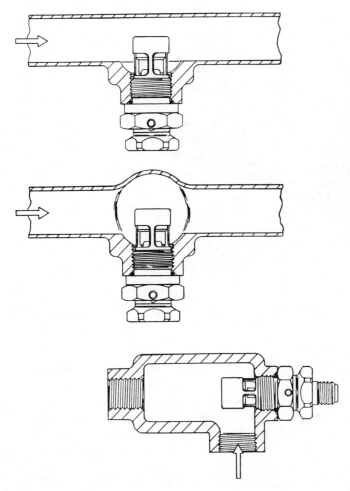

Figure 11.31 Chip collector installations. (Courtesy of Technical Development Company, Glenolden, Pa.)

given interval, such as rate of collection or size of particle, identifies a potential problem. The particles can be analyzed by electron spectroscopy to determine what elements are present and yield clues as to which components are deteriorating [7].

Because the particles trapped are conductors the magnetic probe can be designed such that a gap is bridged between two electrodes when sufficient particles land on the sensor. When this occurs a warning light can be activated. Figure 11.32 shows two such electric chip detectors with radial and axial gaps.

Figure 11.32 Electric chip detector. (Courtesy of Technical Development Company, Glenolden, Pa.)

This type of detector is widely used in aircraft and turbine engine transmissions. Some development of the proper gap size is required since too small a gap will lead to many "nuisance" indications and too large a gap will be insensitive to failure indications. Gaps used are on the order of 0.050 to 0.150 in. With chip collectors or detectors care must be taken to locate them in an area where they will be exposed to as many particles as possible. Also, the area should be relatively still, so the particles can settle.

In forced-flow lubrication systems where particles tend to be dragged along with the oil system, full-flow debris detectors are sometimes used. These configurations have screens through which the total scavenge oil flows. The screens retain the debris. A chip detector can be incorporated for indication purposes. Figure 11.33 shows a full-flow debris detector. An advantage of this type of monitor is that nonmagnetic as well as magnetic debris will be trapped.

Another type of full-flow monitoring device is the indicating screen (Figure 11.34). It is woven from wire strands and when a conductive particle bridges the gap between wires, a warning light can be activated. This device can be used as a pump inlet screen. The indicating screen is sensitive to magnetic and nonmagnetic debris but is more difficult to remove and inspect than a chip detector.

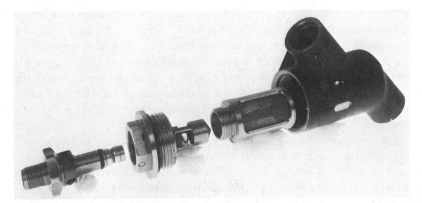

Figure 11.33 Full-flow debris collector. (Courtesy of Technical Development Company, Glenolden, Pa.)

Oil Pressure and Temperature

Oil pressure and temperature measurements are relatively easy to accomplish and yield significant information concerning the operating condition of a gear unit.

The oil pressure should be monitored at the entrance to the unit downstream of any pressure regulation device. Low oil pressure can indicate internal leakage such as a cut static seal or possibly low flow due to pump distress. This can lead to oil starvation of components; therefore, low oil pressure must be

Figure 11.34 Indicating screen debris monitor. (Courtesy of Technical Development Company, Glenolden, Pa.)

investigated and corrected. High oil in pressure can be an indication of downstream blockage such as a clogged oil jet and therefore must also be investigated and corrected. A fluctuating oil pressure can be an indication of pump cavitation due to poor inlet conditions such as excessive pressure drop or air leakage in the pump inlet line. When setting oil pressure limits it must be remembered that the pressure will be affected by variables such as oil temperature and viscosity and jet size variation downstream of the pressure measurement. Pressure in the gearbox cavity will also directly affect the feed pressure.

Oil temperatures should be monitored at the gearbox inlet and outlet at a minimum. The change in temperature across the unit is a good indication of gearbox condition. Scavenge oil temperature of individual bearings can be monitored. On journal bearings, temperature sensors can be embedded under the surface of the babbitt at a depth of 0.030 to 0.060 in. This is a very positive way of monitoring bearing condition. More than one sensor should be installed in case of sensor failure.

Two types of temperature sensors are available: the resistance temperature detector (RTD) and the thermocouple. The RTD is basically a precision resistor where the resistance changes linearly with temperature. RTDs are superior to thermocouples in terms of accuracy, stability, and interchangeability.

REFERENCES

1. AGMA Sound Manual 299.01, Sec. I, Fundamentals of Sound as Related to Gears, American Gear Manufacturers Association, Arlington, Va., May 1978.
2. AGMA Standard 295.04, Specification for Measurement of Sound on High Speed Helical Gear Units, American Gear Manufacturers Association, Arlington, Va., April 1977.
3. AGMA Sound Manual 299.01, Sec. II, Sources, Specifications and Levels of Gear Sound, American Gear Manufacturers Association, Arlington, Va., March 1980.
4. AGMA Standard 426.01, Specification for Measurement of Lateral Vibration on High Speed Helical and Herringbone Gear Units, American Gear Manufacturers Association, Arlington, Va., April 1972.
5. API Standard 670, Noncontacting Vibration and Axial Position Monitoring System, American Petroleum Institute, Washington, D.C., June 1976.
6. Bently Nevada Application Notes, Bently Nevada Corp., Minden, Nev. various dates.
7. Tauber, T., A Design Guide to Effective Debris Monitoring in Gas Turbine Engines and Helicopter Transmissions, Technical Development Company, Glenolden, Pa.

12

MAINTENANCE AND FAILURE ANALYSIS: SCHEDULED MAINTENANCE ACTIONS

The objective of a maintenance program is to ensure satisfactory gearbox performance at all times and to maintain the transmission in a state of readiness if it is not in operation. A program should be planned which includes regular maintenance actions and periodic monitoring of operating characteristics to determine whether any problems are developing.

Every gearbox should come with a set of maintenance instructions developed by the manufacturer. These instructions should include the following information:

General description of the equipment
Specifications as to speed and power ratings, lubrication, dimensions, and so on
Installation information
Lubrication instructions
Maintenance requirements
Assembly and disassembly instructions
Drawing and parts list

The following section presents some general comments on maintenance of gearboxes. A specific unit may require special maintenance procedures and the manufacturer's instructions should be adhered to.

There are several maintenance actions that should be affected during the initial operation of a gearbox. After approximately 50 hr of running time, the following should be accomplished:

Check coupling alignment and correct if necessary.
Check bolt torques and retighten if necessary.

Table 12.1 Regular Gearbox Maintenance Actions

Frequency	Maintenance item	Corrective action
Daily	Check oil temperature and pressure at operating conditions.	If there is a drastic change from previous readings, stop unit and determine cause.
	Check for noise, vibration, and oil leaks.	
	Check sump oil level.	Add oil if necessary.
Weekly	Check oil filter.	Change filter element if necessary.
Monthly	Check lubricating oil for contamination.	Drain and refill lube system if necessary. Change oil filter.
	Check all gages, controls, and alarm systems.	
	Clean breather element.	
Every 2500[a] hours or 6 months	Change lubricating system oil.	

[a]If operating conditions are unusually severe, such as high-temperature or high-moisture atmospheres, oil change requirements might be more frequent. Changes can be based on inspection of the oil for viscosity or acid number in such cases.
Source: Sawyers Turbomachinery Maintenance Handbook, 1st ed., Vol. III, Turbomachinery International Publications, Norwalk, Conn., 1980.

Check piping connections and retighten if necessary.
Change oil and clean sump.

The oil, after 50 hr, need not be discarded but can be drained, filtered through an element with a micrometer rating no greater than the gearbox filter, and reused. Particles may be found in the oil and the sump due to normal wearing in processes. At this point the sump or reservoir should be thoroughly cleaned. After draining the original oil it is recommended that the gearbox and lubrication system be flushed out with a flushing oil. If possible, bring the unit up to operating speed at light load after filling with flushing oil. Shut down immediately after achieving full speed. The drain the flushing oil and refill with the recommended lubricant to the proper level.

After the 50 hr maintenance a regular program should be followed as outlined in Table 12.1. Logs should be kept of instrument readings and maintenance actions to keep a running account of gearbox condition.

When performing maintenance operations every precaution must be taken to prevent foreign matter from entering the gearbox. The introduction of moisture, dirt, or fumes can lead to sludge formation and deterioration of the lubrication and cooling system.

STORAGE

Quite often the gearbox is delivered before the complete system is ready for assembly and it must be stored for some period of time prior to operation. When operation is delayed more than 1 month after shipment, special precautions must be taken to prevent rusting of the components. If possible, the gearbox should be completely filled with oil during storage. Where this is not practical, all exposed metal parts, both inside and outside the unit, should be sprayed with a heavy-duty rust preventative. The gearbox should be stored in a dry area remaining at approximately constant temperature, preferably indoors. If stored outdoors, the gearbox should be raised off the ground and completely enclosed by a protective covering such as a tarpaulin. If possible, the unit should be rotated at weekly intervals while in storage.

OVERHAUL AND SPARE PARTS

Generally, gearboxes do not have a specific time period after which the unit is disassembled and overhauled. It is more common to observe deterioration of components such as bearings and gears during operation and replace the particular component at a convenient time. Usually, the gearbox is delivered with an operating and maintenance manual which describes how to disassemble and assemble the unit. If the user is not completely familiar with the equipment, it would be prudent to have a factory representative accomplish any major component replacements. Spare gears or bearings for gearboxes are not necessarily readily available from the manufacturers. Journal bearings, unlike antifriction bearings, are not usually stocked by distributors. In many cases the bearings are customized for the specific gearbox and therefore are even harder to replace. Finished gears are rarely stocked by manufacturers, and lead times on gears or bearings might be 20 weeks or more. When purchasing the drive the user should request a recommended spares list and determine what the availability of these parts will be. The user and manufacturer can then arrive at some agreement over what spares will be available and where they will be stored.

TROUBLESHOOTING

The major causes of gearbox failure are improper lubrication and overload. Care must be taken to check for proper oil level before operation. Excessive oil

Table 12.2   Gearbox Troubleshooting Chart

Problem	Recommended inspection	Corrective action
Overheating	1. Oil cooler operation	Check flow of coolant and oil flow. Measure oil temperature into and out of cooler. Check cooler internally for buildup of deposits from coolant water.
	2. Is oil level too low or too high?	Check oil level indicator.
	3. Bearing installation	Make sure that bearings are not pinched and properly adjusted.
	4. Grade and condition of oil	Check that oil is specified grade. Inspect oil to see if it is oxidized, dirty, or with high sludge content.
	5. Lubrication system	Check operation of oil pump. Make sure that suction side is not sucking air. Measure flow. Check if oil passages are free. Inspect oil line pressure regulator, nozzles, and filters to be sure they are free of obstruction.
	6. Coupling float and alignment	Check coupling alignment and adjust end float.
Shaft failure	1. Type of coupling	Rigid couplings between rigidly supported shafts can cause shaft failure. Replace with coupling to provide required flexibility and lateral float.
	2. Coupling alignment	Realign equipment as required.
	3. Excessive overhung load	Reduce overhung load. Use outboard bearing or replace with higher capacity unit.
	4. High transient loading	Apply couplings capable of absorbing shocks. Use couplings with shear pins.

Table 12.2 (Continued)

Problem	Recommended inspection	Corrective action
Shaft failure (continued)	5. Torsional or lateral vibrations	Adjust system mass elastic characteristics to control critical speed location. Possibly, coupling geometry can be modified.
	6. Cracks due to fretting corrosions	Note cause of fretting and correct. Press fits between gear and shaft.
Oil leakage	1. Exceed oil level	Check oil level indicator.
	2. Is breather open	Check oil breather.
	3. Are oil drains open	Check that all oil drain locations are free and clean.
	4. Oil seals	Check oil seals and replace if worn. Check condition of shaft under seal and polish if necessary.
	5. Plugs at drains, levels, and pipe fittings	Apply sealant and tighten fittings.
	6. Housings and caps	Tighten cap screws or bolts. If not effective, remove housing cover and caps. Clean mating surfaces and apply new sealing compound. Reassemble. Check compression joints by tightening fasteners firmly.

Source: *Sawyers Turbomachinery Maintenance Handbook*, 1st ed., Vol. III, Turbomachinery International Publications, Norwalk, Conn., 1980.

volume can be as detrimental as lack of lubrication and will result in churning and overtemperature of components. Overload can be a result of vibration, shock loads, or high torque at low speed. If there is a possibility that operating loads will exceed rated gearbox loads, the manufacturer should be consulted. Table 12.2 is a troubleshooting guide that gives some guidance in how to identify and correct some of the common problem areas occurring during gearbox operation.

Gear teeth should be inspected for nicks, burrs, and scratches, which may be repaired by blending provided that they are minor and not on the working surfaces of a tooth. The blend may be accomplished using a small file and an India or carborundum stone. Crocus cloth should be used for the final polishing. All repairs must be finished smoothly. Power tools are not permitted for blend repairs.

In many gearboxes the teeth can be visually inspected by removing inspection covers bolted into the casing. When opening these inspection covers care must be taken to ensure that no foreign material enters the gearbox. Gear teeth should be examined under good lighting and be wiped clean of oil to prevent a false diagnosis. The content pattern should cover approximately 80% of the tooth. The following section discusses gear tooth failure modes and describes various conditions that may be encountered in the field.

Gear Tooth Failure Modes

The major modes of gear tooth failure are breakage, wear, pitting, and scoring. When these problems are encountered in the field it is important to accurately define the condition and causes of failure to be able to determine corrective actions for the particular units in difficulty and also to modify analytical and manufacturing methods for future gearboxes so that they will not suffer the same problems. Also, an accurate diagnosis of field problems will enable the user to determine if a gearbox requires immediate modification or replacement or if it can be expected to continue to function for some period of time, when repairs can be made more conveniently.

Breakage

Breakage of gear teeth is the most catastrophic form of failure. It occurs precipitously with no advance warning. If a number of teeth break, load transmission is no longer possible. If only one tooth or a portion of a tooth breaks, there is the possibility of secondary damage if the broken part interferes with other components in the system. Also, dynamic loading will increase and if operation continues, other teeth will soon fail. Breakage of gear teeth is caused by excessive bending stress in the root imposed by the transmitted load. Tooth breakage can be the result of a fatigue mechanism (Figure 12.1) or an overload which exceeds the gear tooth fracture strength (Figure 12.2). A fatigue break initiates as a small crack which, over a large number of load cycles, propagates until a portion of, or a whole tooth, separates from the gear. Failures of this nature may be a result of system overloads greater than the design load, such as torsional vibrations. Manufacturing discrepancies such as tool marks or metallurgical discrepancies can also lead to fatigue failures by initiating cracks. In surface-hardened gears, if the case in the root is not correct either due to heat-treat problems or excessive machining, bending failures can result.

Figure 12.1 Fatigue breakage. (Courtesy of American Gear Manufacturers Association, Arlington, Va.)

As shown in Figure 12.1, the smooth appearance of the fracture surface attests to the fact that considerable working of the cracked surfaces occurred prior to final separation. Overload breakage occurs in relatively few cycles; therefore, the fracture surfaces are rougher than those of a high-cycle fatigue failure. Some causes of overload breakage are:

Large particles passing through the mesh
Sudden misalignments such as when a coupling fails
Bearing seizures
Shock loads such as short circuits in a generator drive

Figure 12.2 Overload breakage. (Courtesy of American Gear Manufacturers Association, Arlington, Va.)

In some cases other failure modes such as wear or pitting may increase dynamic loading or weaken the tooth to the extent that breakage ultimately occurs.

Wear

Wear may be defined as the loss of metal due to the rubbing action of two surfaces moving in relation to one another when the oil film is not of sufficient thickness to separate them [1]. One form of rubbing wear is adhesive wear characterized by metal particles from one gear tooth adhering to the mating gear tooth by a welding action and subsequently detaching. The other form of rubbing wear is abrasive wear caused by abrasive action between the sliding gear teeth or by the presence of abrasive particles between them. These particles may be contaminants in the oil or pieces detached from the tooth surfaces themselves. Figure 12.3 illustrates some worn areas on a helical gear.

Gear teeth do not necessarily wear during operation. If the oil film thickness is sufficient to separate the mating tooth surfaces, millions of cycles can be accumulated with no measurable wear. In some cases there may be an initial wearing in of gear teeth and if rubbing wear diminishes with time it may not be detrimental; therefore, when wear is first noted the gearset should be closely monitored to determine the rate at which wear progresses such that a determination can be made as to whether it will be damaging. The wear-in

Figure 12.3 Tooth wear.

Figure 12.4 Initial pitting. (Courtesy of American Gear Manufacturers Association, Arlington, Va.)

phenomenon is more common with through-hardened gears (Rc 32 approximately) than surface-hardened gearing (approximately Rc 60).

Pitting

Pitting manifests itself in several forms. Initial pitting, sometimes called frosting, may occur during early operation. This is a surface-oriented failure mode where local high spots contact and exhibit distress, as shown in Figure 12.4. As the high spots are removed, the load is more evenly distributed and the pitting action diminishes. This condition may be acceptable.

Destructive fatigue pitting is a result of repeated stress cycling of the tooth surface beyond the material's endurance limit. Surface or subsurface cracks initiate, propagate, and eventually material detaches from the tooth surface, leaving pitted areas (Figure 12.5). Pitting may progress to a point where large areas are broken out, as shown in Figure 12.6. This condition is referred to as spalling.

Scoring

Scoring is a form of surface damage on the tooth flank which occurs when overheating causes the lubricant film to become unstable, allowing metal-to-metal contact. Local welding is initiated and the welded junctions are torn apart by the relative motion of the gear teeth, resulting in radial score marks. Figures 12.7 to 12.9 illustrate degrees of the scoring phenomenon. Light scoring which does not progress may be acceptable, but heavier scoring can destroy the tooth profile and lead to pitting and breakage. The scoring type of failure mode generally occurs in high-speed applications using low-viscosity lubricants.

Figure 12.5 Pitting.

Evaluation of Surface Distress

Tooth breakage is the only failure mode that demands immediate action be taken. Wear, pitting, and scoring are progressive and may occur over a long period of time without significantly impairing the function of the unit. As an example, Figures 12.10 to 12.16 show the progressive deterioration of a gear during 1000 hr of operation. At 50 hr there is a band of wear in the dedendum of the teeth with a depth of 0.0001 in. at the right hand end increasing to 0.0006 in. at the more heavily worn left-hand end. Some pitting is also noticeable. At 120 hr the wear has spread toward the tip of the tooth. More pits have appeared. At 200 hr the tooth tip has worn further, to a depth of approximately 0.0006 in. At 400 hr the wear band has spread to the right-hand end. During this segment of testing the right-hand end wore 0.0006 in. From 400 to 1000 hr there were no major changes in gear tooth condition.

The purpose of this discussion is not to advocate running gears in a worn and pitted condition. The system noise and vibration level will increase as the tooth surface deteriorates and abrasive particles are being introduced into the lubrication system. There is always the danger of catastrophic breakage of a weakened tooth. It can be seen, however, that even when the tooth surface is severely deteriorated, gear teeth are capable of operating.

As a general guideline, if surface distress is noted early in operation and it is light in nature, time can be taken to observe the progression and if distress

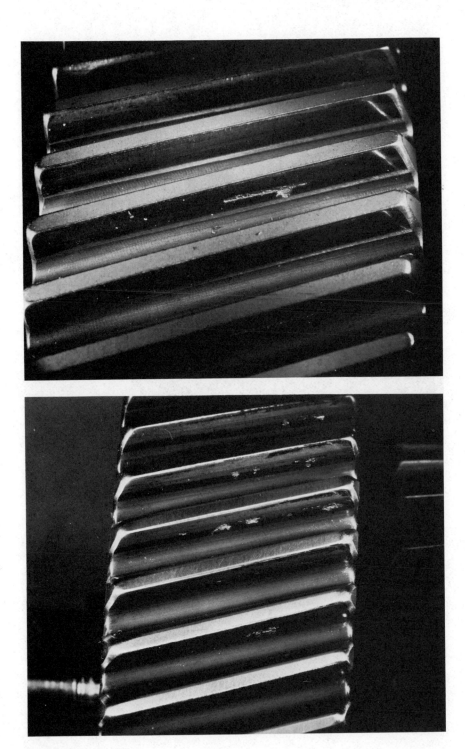

Figure 12.6 Spalling.

Figure 12.7 Light scoring.

Figure 12.8 Medium scoring.

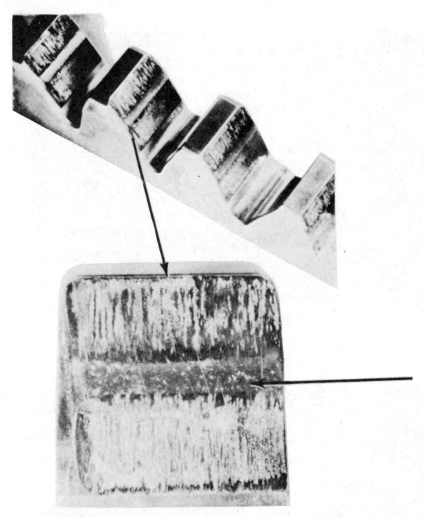

Figure 12.9 Heavy scoring.

ceases to increase, the gearset is probably satisfactory for continued operation. If distress occurs after long operation, it is probably an indication that something has changed in the operating conditions. Possibly loads have increased or the lubricant has deteriorated. When surface distress appears there are several potential changes in operating procedures that can be made to prolong gear life:

It may be possible to reduce the load on the gear teeth. Sources of external loading such as coupling unbalance or misalignment should be checked. If

Figure 12.10 Tooth condition after 50 hr.

Figure 12.11 Tooth condition after 120 hr.

Figure 12.12 Tooth condition after 200 hr.

torsional vibrations are suspected, measurements should be taken. It may
be possible to derate the system and operate at lower loads.
An oil with higher load capacity may be available. EP additives can possibly be
of value. The manufacturer should be consulted before changing lubricants.
Operating the unit with cooler oil which will be at a higher viscosity creating a
thicker film may be beneficial when surface distress is a problem.

Figure 12.13 Tooth condition after 400 hr.

Figure 12.14 Tooth condition after 600 hr.

Figure 12.15 Tooth condition after 800 hr.

Figure 12.16 Tooth condition after 1000 hr.

Cracking

On occasion, cracks are observed on gear teeth as shown in Figure 12.17. Cracks are usually a result of incorrect heat treatment or abusive grinding. In some cases improper heat treatment results in brittle tooth tips or ends that chip off during operation.

Cracks on the tooth flanks are dangerous since they can propagate and lead to catastrophic failure; therefore, gears exhibiting cracking should be replaced. Chipping of gear teeth at the tips or ends may be repairable by blending. The danger here is that if more chipping occurs, the pieces may interfere with rotating components.

Improper heat treatment of case-hardened gears can result in case-core separation. Cracks originating in the core material propagate along the case-to-core boundary and then work out to the surface. When this occurs large areas of material are removed.

Oil Starvation

Quite a few gearboxes have been lost because the units were run without lubrication. Sometimes when viewing the components after failure it is not obvious what the cause was. For instance, Figure 12.18 shows the condition of components of a planetary gear set after 1.75 min of operation at full load and speed without oil. The smaller sun gear has failed catastrophically, shedding all its teeth. Other gears and antifriction bearings are in reasonably good condition. The sun gear, having the least area to dissipate heat, reached the melting point

Figure 12.17 Tooth cracking.

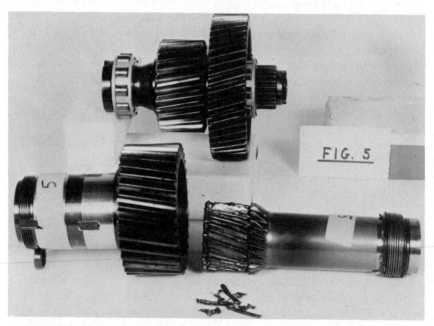

Figure 12.18 Oil starvation failure. (Courtesy of AVCO, Lycoming Corporation, Stratford, Conn.)

of its teeth and suffered the most damage. It was not obvious from the component condition that oil starvation was the cause of failure. With journal bearing gearboxes the diagnosis is easier since the bearings will sieze long before the melting point of gear teeth is reached.

REFERENCE

1. Southwest Research Institue, Gear Tooth Scoring Investigation, Report USAAMRDL-TR-75-33, San Antonio, Tex., July 1975.

APPENDIX

INVOLUTE TABLES (INV ϕ = tan ϕ - ϕ)

Angle (deg)	Cosine	Involute	Angle (deg)	Cosine	Involute
17.0	0.956305	0.009025	19.2	0.944376	0.013134
17.1	0.955793	0.009189	19.3	0.943801	0.013346
17.2	0.955278	0.009355	19.4	0.943223	0.013562
17.3	0.954761	0.009523	19.5	0.942641	0.013779
17.4	0.954240	0.009694	19.6	0.942057	0.013999
17.5	0.953717	0.009866	19.7	0.941471	0.014222
17.6	0.953191	0.010041	19.8	0.940881	0.014447
17.7	0.952661	0.010217	19.9	0.940288	0.014674
17.8	0.952129	0.010396	20.0	0.939693	0.014904
17.9	0.951594	0.010577	20.1	0.939094	0.015137
18.0	0.951057	0.010760	20.2	0.938493	0.015372
18.1	0.950516	0.010946	20.3	0.937889	0.015609
18.2	0.949972	0.011133	20.4	0.937282	0.015850
18.3	0.949425	0.011323	20.5	0.936672	0.016092
18.4	0.948876	0.011515	20.6	0.936060	0.016337
18.5	0.948324	0.011709	20.7	0.935444	0.016585
18.6	0.947768	0.011906	20.8	0.934826	0.016836
18.7	0.947210	0.012105	20.9	0.934204	0.017089
18.8	0.946649	0.012306	21.0	0.933580	0.017345
18.9	0.946085	0.012509	21.1	0.932954	0.017603
19.0	0.945519	0.012715	21.2	0.932324	0.017865
19.1	0.944949	0.012923	21.3	0.931691	0.018129

Angle (deg)	Cosine	Involute	Angle (deg)	Cosine	Involute
21.4	0.931056	0.018395	25.3	0.904083	0.031130
21.5	0.930418	0.018665	25.4	0.903335	0.031521
21.6	0.929776	0.018937	25.5	0.902585	0.031917
21.7	0.929133	0.019212	25.6	0.901833	0.032315
21.8	0.928486	0.019490	25.7	0.901077	0.032718
21.9	0.927836	0.019770	25.8	0.900319	0.033124
22.0	0.927184	0.020054	25.9	0.899558	0.033534
22.1	0.026529	0.020340	26.0	0.898794	0.033947
22.2	0.925871	0.020629	26.1	0.898028	0.034364
22.3	0.925210	0.020921	26.2	0.897258	0.034785
22.4	0.924546	0.021216	26.3	0.896486	0.035209
22.5	0.923880	0.021514	26.4	0.895712	0.035637
22.6	0.923210	0.021815	26.5	0.894934	0.036069
22.7	0.922538	0.022119	26.6	0.894154	0.036505
22.8	0.921863	0.022426	26.7	0.893371	0.036945
22.9	0.921185	0.022736	26.8	0.892586	0.037388
23.0	0.920505	0.023049	26.9	0.891798	0.037835
23.1	0.919821	0.023365	27.0	0.891007	0.038287
23.2	0.919135	0.023684	27.1	0.890213	0.038742
23.3	0.918446	0.024006	27.2	0.889416	0.039201
20.4	0.917755	0.024332	27.3	0.888617	0.039664
20.5	0.917060	0.024660	27.4	0.887815	0.040131
20.6	0.916363	0.024992	27.5	0.887011	0.040602
20.7	0.915663	0.025326	27.6	0.886204	0.041076
20.8	0.914960	0.025664	27.7	0.885394	0.041556
20.9	0.914254	0.026005	27.8	0.884581	0.042039
24.0	0.913545	0.026350	27.9	0.883766	0.042526
24.1	0.912834	0.026697	28.0	0.882948	0.043017
24.2	0.912120	0.027048	28.1	0.882127	0.043513
24.3	0.911403	0.027402	28.2	0.881303	0.044012
24.4	0.910684	0.027760	28.3	0.880477	0.044516
24.5	0.909961	0.028121	28.4	0.879649	0.045024
24.6	0.909236	0.028485	28.5	0.878817	0.045537
24.7	0.908508	0.028852	28.6	0.877983	0.046054
24.8	0.907777	0.029223	28.7	0.877146	0.046575
24.9	0.907044	0.029598	28.8	0.876307	0.047100
25.0	0.906308	0.029975	28.9	0.875465	0.047630
25.1	0.905569	0.030357	29.0	0.874620	0.048164
25.2	0.904827	0.030741	29.1	0.873772	0.048702

Angle (deg)	Cosine	Involute	Angle (deg)	Cosine	Involute
29.2	0.872922	0.049245	33.1	0.838671	0.074188
29.3	0.872069	0.049792	33.2	0.836764	0.074932
29.4	0.871214	0.050344	33.3	0.835807	0.075683
29.5	0.870356	0.050901	33.4	0.834848	0.076439
29.6	0.869495	0.051462	33.5	0.833886	0.077200
29.7	0.868632	0.052027	33.6	0.832921	0.077968
29.8	0.867765	0.052597	33.7	0.831954	0.078741
29.9	0.866897	0.053172	33.8	0.830984	0.079520
30.0	0.866025	0.053751	33.9	0.830012	₁0.080305
30.1	0.865151	0.054336	34.0	0.829038	0.081097
30.2	0.864275	0.054924	34.1	0.828060	0.081894
30.3	0.863396	0.055518	34.2	0.827081	0.082697
30.4	0.862514	0.056116	34.3	0.826098	0.083506
30.5	0.861629	0.056720	34.4	0.825113	0.084321
30.6	0.860742	0.057328	34.5	0.824126	0.085142
30.7	0.859852	0.057940	34.6	0.823136	0.085970
30.8	0.858960	0.058558	34.7	0.822144	0.086804
30.9	0.858065	0.059181	34.8	0.821149	0.087644
31.0	0.857167	0.059809	34.9	0.820152	0.088490
31.1	0.856267	0.060441	35.0	0.819152	0.089342
31.2	0.855364	0.061079	35.1	0.818150	0.090201
31.3	0.854459	0.061721	35.2	0.817145	0.091066
31.4	0.853551	0.062369	35.3	0.816138	0.091938
31.5	0.852640	0.063022	35.4	0.815128	0.092816
31.6	0.851727	0.063680	35.5	0.814116	0.093701
31.7	0.850811	0.064343	35.6	0.813101	0.094592
31.8	0.849893	0.065012	35.7	0.812084	0.095490
31.9	0.848972	0.065685	35.8	0.811064	0.096395
32.0	0.848048	0.066364	35.9	0.810042	0.097306
32.1	0.847122	0.067048	36.0	0.809017	0.098224
32.2	0.846193	0.067738	36.1	0.807990	0.099149
32.3	0.845262	0.068432	36.2	0.806960	0.100080
32.4	0.844328	0.069133	36.3	0.805928	0.101019
32.5	0.843391	0.069838	36.4	0.804894	0.101964
32.6	0.842452	0.070549	36.5	0.803857	0.102916
32.7	0.841511	0.071266	36.6	0.802817	0.103875
32.8	0.840567	0.071988	36.7	0.801776	0.104841
32.9	0.839620	0.072716	36.8	0.800731	0.105814
33.0	0.838671	0.073449	36.9	0.799685	0.106795

INDEX